The Pensado Papers

The Pensado Papers

The Rise of Visionary Online Television Sensation *Pensado's Place*

DAVE PENSADO *and* HERB TRAWICK
with MAUREEN DRONEY

HAL LEONARD BOOKS
An Imprint of Hal Leonard Corporation

Published in 2014 by Hal Leonard Books
An Imprint of Hal Leonard Corporation
7777 West Bluemound Road
Milwaukee, WI 53213

Trade Book Division Editorial Offices
33 Plymouth St., Montclair, NJ 07042

Printed in the United States of America

Front cover design by Tilman Reitzle | Oxygen Design
Front cover photograph by Zan Nakari | ZanPhotos.com
All other photography by Brian A. Petersen
Book design by Kristina Rolander

Library of Congress Cataloging-in-Publication Data
Pensado, Dave.
The Pensado papers : the rise of visionary online television sensation, Pensado's place / Dave Pensado and Herb Trawick with Maureen Droney.
pages cm
Includes index.

ISBN 978-1-4803-4569-0

1. Pensado, Dave. 2. Sound engineers--United States. 3. Popular music--Production and direction. I. Trawick, Herb. II. Droney, Maureen. III. Title.
ML429.P405A5 2014
781.49092--dc23
2014025461

www.halleonardbooks.com

To Joy and Melissa Pensado,
for their unconditional love and unwavering support
during the good times and bad. —Dave

This book and all that I do—great or small—
will always and forever be dedicated to my greatest love,
my inspiration, the person who most advocated for me
to be me, my heart and soul, Toni Elizabeth Scott.
Your memory is my reason for being. —Herb

CONTENTS

PREFACE

My phone rang at an unusual time. It was Neal Pogue, star mixer, famed for his work with the hit band Outkast, saying, “I can’t believe what’s going on with Dave!”

I was startled. How the heck did Neal know that Dave had just stood me up for a meeting—hadn’t even called—leaving that fine Angus beef burger he’d said he wanted to go cold? Were all the engineers sharing calendars? Boy, was I wrong—way wrong.

Neal continued, “Tell me what’s happening. I heard he might not make it.” *What??* After collecting my senses I asked Neal to start from the beginning, and the facts began to emerge. Dave had suffered a stroke—a serious, life-threatening stroke. He was in a medically induced coma and would remain that way for 11 days as the doctors fought to keep him alive. I tracked down Joy, his wife, and the picture she painted of his condition—the breathing tube, the doctors and nurses on 24-hour watch, the prognosis, virtually everything—pointed to a very unhappy ending.

I was gobsmacked. Shattered. Rendered completely helpless with no move that I could make. Dave had been at my place a lot lately, asking me to think about his career. That’s why we’d planned to meet that day. I wasn’t his manager at the time; I was just his friend. A blood brother, truth be told. And now I was about to lose him.

Ninety days later, I was the first person Dave spoke with outside of his immediate family. His speech was slurred, almost unintelligible, and Joy had warned me he was a bit delusional. He said, “Herb, I’m okay.” He wanted to make sure that we would meet when he got out of the hospital. He told me he loved me, and I told him I loved him back. Then I hung up, pulled over into a parking lot in Encino, California, and cried like a baby.

This is a story of incredible, almost unbelievable redemption—of Dave's journey from death's door to being the best he's ever been. It's a story of one of the biggest hearts ever and of a partnership strengthened by trust, perseverance, and new horizons. It's also a whole lot of fun, with teachings from our friends, engineering, philosophy, a look behind our show, and news you can use. It's the whole shebang. These are the *Pensado Papers.*

—Herb Trawick

ACKNOWLEDGMENTS

DAVE PENSADO'S ACKNOWLEDGMENTS

Special thanks to:

Darlene, my sister, whose prayers God answers.

Rick Tarkington, my friend and brother. In college, touring Europe, in our first band, and in life itself, you've been a true compadre.

Todd Chapman. We made some great records and caught a lot of fish. Thanks for your friendship.

Larry Turner, band mate, friend, and the main reason I'm an engineer.

Phil Benton. Thank you for my first gig.

Ed Seay. You gave me the encouragement and confidence to succeed.

Richard Wolf and Bret "Epic" Mazur. We made my favorite records. I owe you guys so much.

Jac Colman. You have always believed in me, and your dedication to the arts is truly a blessing and an education to me.

Paul Davis. I miss you, Paul. You cared. Still the most gifted artist I've ever known.

Bruce Sonneborn, Craig Burbidge, and Keith Andes. Thank you.

All of my assistants, especially Dylan, Jaycen, Ariel, Andrew, Dave, Anthony, and Ethan Willoughby. I miss you daily, Ethan. You left us way too soon.

To Herb Trawick: I've known you for a couple of decades and I'm still daily in awe of you. You constantly blow me away with what you create and accomplish, from my career in the studio to *Pensado's Place*. You have never once disappointed or failed me. Your intelligence is only eclipsed by your heart. A mere "thank you" is inadequate, so all I can say is love ya, my friend.

—Dave Pensado

HERB TRAWICK'S ACKNOWLEDGMENTS

I would like to give special thanks to:

Herb, Sr., for trailblazing with grace.

Emma Jean, my mother, for demanding that I dream.

Tony Banks for being my little brother.

To Aretha and Tyler for your perseverance and love, and for naming the book.

And, most importantly, thanks to the family: Steve "my brother" Ivory, Fred "my brother" Terrell, Jonelle "my sister" Procope (for neverending support), Ron "my brother" Dixon, Steve "my brother" McKeever (for understanding), Nicole "lil sis" Avant and Ted Sarandos (for loving Nicole and always supporting us), Darrell "my brother" Miller, the Crew, Virg and Brenda, Frazier and Dee, Gil and Val, Harp and Lisa, Walt and Jan, Tarlin and Grace, Tanika, Coach Steve "my mentor" Gilmore, Ron "the man" Dixon, and Rodney "50" Johnson.

Maureen Droney for her diligence and pen.

Donnie Simpson, Will Thompson, Greer Williams, Shevy Shovlin, Team Pensado, Pensado Students, every guest on the show, and our awesome and incredible audience and fans for the gift you have bestowed on us. We are so, so grateful!

Finally, to Dave Pensado: There simply are no words we haven't said—private and public. Your heart, courage, tolerance, patience, trust, belief, encouragement, intellect, and care take my breath away. Fighting beside you in the foxhole of life has been truly affirming. Through whatever we've done—no matter how it's been received—it's the highest honor to do it together. Thanks, DP.

—HT

The Pensado Papers

1 Island (Hip) Hopping

I don't do hip-hop. Oops, I guess I do!
The first big hit.

I wasn't quite sure how or why I'd been called for this particular job. I had been in L.A. about six months, some of that time living in my car—which sucks. I don't recommend it. I had plenty of time on my hands and I spent most of it perfecting my ability to sneak into recording studios—big, small, famous, infamous, or unknown. Palaces, dumps, and dives; as long as it had a recording console, mics, and speakers, I didn't care. This kept me so busy that I didn't realize that by doing this I was actually learning the business, getting the lay of the land and meeting people who would later impact my career. What I did realize was, "Man, I love this stuff!"

At a seminal studio called Skip Saylor Recording, I met this cat named Herb who seemed pretty cool. We spent a couple of hours talking and promised to stay in touch. Little did I know it was a meeting that would end up changing my life. Two months later, Herb referred me to Kevin Fleming, a vice president at Island Records who wanted me to take a shot at a mix for some producer friends. Kevin bottom-lined it to me: "It's a hip-hop mix on a really hot act. You need to start tomorrow. Do you want it or not? They need to know right now."

My brain was going: "Huh? Me? Um, I'm not really ready. It's a big act, lots of pressure—I should say no, I must say no. . . ." So of course I said yes. Well, actually, I said, "Hell, yeah!"

The artist was Bell Biv DeVoe and they were coming off a smash album called *Poison*. "Do Me Baby" was the title of the next scheduled single. The producers, Wolf & Epic (a.k.a. Richard Wolf and Bret Mazur), were looking for someone new who could create a fresh version of "Do Me Baby" for a remix release. They wanted something aggressive and in your face, with almost a rock attitude. Wolf & Epic, by the way, were white dudes who shared my philosophy, not only about embracing racial diversity, but also about being wholly comfortable living in it. Bell Biv DeVoe were the same. Cool as could be, ahead of the curve, and they knew they were onto something. We were all in the same place, and together we went into the lab fired up and ready to go.

I'm not generally the shrinking violet type, but as the newbie my intent was just to stay in my lane. After the first day, that went out the window. We were charged up. No fear, no reservations. In retrospect, the whole thing was a majorly ballsy move by Wolf & Epic. BBD was on the fast track. They were the artist face of New Jack Swing with their album *Poison* headed toward triple platinum, and we could have derailed that if we'd screwed it up. The pressure was huge. What helped us was, we decided to blow up the boundaries instead of allowing ourselves to be constrained by them. We just didn't give a shit. It was go time!

> "If someone pays you a dollar you need to give that person two dollars' worth of work."

Alas, there was an unexpected twist. Our remix of "Do Me Baby" was a bona fide smash. It went to number two on the pop charts, just as the "Poison" single had done. "Smack it, flip it, rub it down, oh no!" became a catch phrase used everywhere. New Jack Swing was hot, and kids were dressing like BBD, dancing like BBD, and talking like BBD. The records worked at urban, pop, R&B, and hip-hop. It was a phenomenon and a full-on lifestyle and culture marker.

Problem was, I wasn't on the rocket ship. Because I had gotten no credit on "Do Me." Zero. Zip. Nada. I hadn't even thought about credits at the time. I was just happy to be working. Very big mistake! Now, let's be clear: nobody screwed me; there was nothing nefarious. I just wasn't aware. Didn't know, didn't have representation; there was a rush—stuff happens. But that was the last time I didn't think about it. It was pretty tough to watch all that success with no mention of my contribution. I thought I'd blown it. Turns out I hadn't—thank the Lord.

Since "Do Me Baby" was so successful, BBD returned to the same well for "Thought It Was Me," and back to the woodshed we went. Rock attitude, hard drums, and two or three E-mu SP 1200s running live—virtual, no drums to tape—made the record less dry than the album version. Plus, credits on the record for everyone—including me. Would it matter? There was no way to know.

Then, boom! The record skyrocketed to number one R&B and went top 20 pop. BBD was killing it all over the airwaves—radio and TV. Wolf & Epic were on fire as producers, and Kevin Fleming at Island Records was looking pretty smart for recommending an unknown Southern yokel. My phone started ringing nonstop. A&R folks who didn't even know me were suddenly my best friends and I was taking lunch meetings all over town. I went from living in my car to eating like a king, and I thought to myself, "Sheiiiiit, this is all right. I gotta keep this going!"

Oh, and something else key happened during the "Do Me Baby" / "Thought It Was Me" period. Wolf & Epic gave me the nickname "HardDrive." I didn't think much of it at the time—the reference was to work ethic, not computer hardware. The name didn't make the first record, but it made the second—and it stuck. It was everywhere in my business life. And the role it played in my personal life turned out to be earth shattering.

2 Go West, Young Mullet

How a guitar player of Spanish descent from Florida and Georgia, who'd gigged with Motown bands, James Brown, and members of Lynyrd Skynrd and the Allman Brothers, ended up in L.A.—and the merriment that ensued.

DAVE: My original impetus to be an engineer didn't come from a burning desire to engineer. It was just that I was living in Atlanta and I'd reached the point where I wanted to stay in music but I didn't want to have to be in bands and travel anymore.

Being in bands taught me a lot (I've played in many). Especially in the South, you learn very quickly that you either entertain the crowd or you die—there's nothing in between. A lot of the time our gigs were like something from a Blues Brothers movie—chicken wire in front of the stage and all. And because of the abject poverty that is the normal state of being a musician, white bands would take on gigs in full R&B venues and black bands would take on gigs in full rock venues. You needed the money too much to be selective, so you just had to figure out a way to get through it.

You learned quickly not to play your best song first, because if you did, you had no place to go. And you learned to play your worst songs toward the end of the set. You learned where to put the ballad to pace the set—right before your best song is usually good! Running your best song fourth or fifth in the set is also usually good.

It's always best to start off with something accessible to get the crowd going in your direction. And, while I know this may not be truly smart, you always try to play as loud as possible—up to the point of getting fired, of course. I admit we did get fired a lot, but the crowd loved it. At one point I even built myself a guitar with no knobs just so nobody could tell me to turn it down.

It was not unusual for us to do our first set at midnight and our last at four, five, or six in the morning. Performing a 45-minute set every hour was great training, because we had to entertain. Our job was to give the audience something they enjoyed, something memorable that enhanced their experience of being out in a club. We had to make them want to be there because the club owner had to make money, at the door and selling drinks.

HERB: I once had the privilege of working with a Cotton Club habitué who called that feeling "Peas and Potatoes." It was Maurice Hines, from a famous Harlem act called Hines, Hines, and Dad, who used the term. When I asked him what it meant, he told me, "The Cotton Club was a dinner club and we did four sets a night. Our father was the taskmaster and he'd tell us, 'If you don't perform something that makes them look up from their peas and potatoes, between sets I'm gonna kick your asses.'"

Because that's when he knew they had it, when everybody in the audience stopped eating to listen. So to Dave's point: When you learn in those kinds of environments, you learn to go full on. You don't know how to do it any other way.

DAVE: For me, morphing all of that performing experience into a studio environment was really just doing the same thing. But instead of trying to please a crowd, I was trying to get the music that came out of the speakers to give me the same feeling I had when I was on stage. I was reaching for the feeling that if a crowd of people were in the room they'd probably like it. And that's what I do to this day.

It's weird how it happens to you. At some point, when you're very young, there's a song that gives you an overwhelming feeling. The groove, the music—something—just consumes you. It takes you over and sort of lifts a gray veil off your eyes. When you get that feeling, you not only want to experience the music, you want to help create it. And the transition from feeling it to creating it is a very interesting process. Because at that point, you're not trying to earn a living when you create it—you're just trying to create it.

In those early band days, you're living at home; you don't have a lot of expenses. You're playing for your friends and anybody who wants to listen, striving for the feeling. It's music at its purest. The pursuit of that feeling is what keeps me going, even today. It's what every engineer strives for. They're sitting in a studio trying to get that feeling by listening to what's coming out of a set of speakers.

Each guest on *Pensado's Place* might say it differently, but it's something that permeates throughout all our conversations. No matter what field of audio they're in, no matter what genre of music, what did it for them and got them going in this world is the feeling that they got when they heard that first important record.

Every time you sit down to mix, you're trying to get back to that place. I'm resisting the temptation to compare all this to heroin. But the analogy is there, in the sense that the first time you do a drug it's the most incredible experience, the second time it's not as good, and the third time is worse still. What causes people to become addicted is the quest to have that that first feeling over and over again. Sorry to use that metaphor, but it sure is apropos to this situation!

INFLUENCES AND INSPIRATIONS: STARTING IN THE SOUTH

DAVE: Because my family is Spanish, and before we moved to Georgia we lived in South Florida, I grew up with Latin music. A lot of it was Cuban. The artist Cachao, for example, was a big influence on me. And my mother was very musical; she played guitar and taught me. Her teaching style was not "hit your knuckles with a ruler." It was more that I'd just snatch the guitar away from her and play. Until I got my own guitar we fought more over the guitar than anything.

Mom also loved math, the practical application of mathematics, and we'd do algebra equations, just for fun, when I was three and four years old. Then we'd sit out in the yard and she'd tell me the names of all the bugs we'd run across. My mom and dad both imbued me with curiosity about the world around me. Of all the gifts I got from them, I think that's my favorite. Because of that gift, I have a curiosity about everything around me. Like, sitting at this table right now, I'm trying to figure out the names of the flowers in the vase on the table, and the material the vase is made out of, and I'm wondering about the finish on the table. Is it lacquer, or urethane? I'm curious about everything around me.

From my father I also got this wonderful—by my measure anyway—sense of sarcasm and smart-assedness, along with total skepticism regarding everything around me. I rarely believe anything I'm told because of my dad, who questioned everything.

"You teach people how to treat you."

In the South I played in a couple of successful bands, and I played with great musicians—some of them well known, and some of them not, but still so good I've not run into people more talented anywhere. But there comes a point when you want to do more.

I'd learned a lot about audio by doing live sound. For me, it was always about the quality of the music itself.

Engineering was just a way to amplify the musicality, to help create the emotion, the vibe, and the feel. As it turns out, Atlanta was the perfect place to learn and to grow as an engineer because there were so many wonderful people there who guided and helped me.

There was Phil Benton, who had produced the Brick albums and also engineered their early big records. Phil was a top-notch producer. Then Ed Seay came along. He listened to my work and gave me some credibility and confidence. There were three studios in Atlanta that were very important, and that I worked at: Bob Richardson's Mastersound, Tom Wright's Cheshire Sound, and Eileen West's Web IV. At Web IV a lot of great recordings were in the tape vault. Of course I'd sneak in there, pull them out, listen, and try my hand at remixing! A lot of Van Morrison tapes were back there, too, and I did the same with them. It was one of the premier studios in Atlanta at the time. Then Tom Wright took me under his wing at Cheshire Studios, where L.A. Reid and Babyface worked when they were getting their start. Prince used to work there, too.

LEARNING BY DOING

DAVE: I was never an assistant engineer. Instead, in Atlanta, I had a number of happy accidents.

One happy accident was that I embarked on my engineering career with guitar player Larry Turner, one of my closest friends, with whom I was in many bands. We started together, and we became known as an engineering team. Combined, we didn't equal one whole engineer, but we were learning as we went. When Larry was lead, I would be setting up mikes and helping him. When I was point, he'd help me. Sometimes he'd get tired and I'd take over, and vice versa. We were sharing experiences and learning together—just absorbing it all. When we had down time we were always recording something—anything from bird calls outside to gunshots inside, firing .45s into phone books. One day we set out to replace a 12-inch speaker from a guitar cabinet that wasn't working well. Once it was out, we decided to lay it on top of a snare drum and run a signal from the recorded snare back out to that speaker. We re-miked that and got a little more ambience to add to our snare sound—it was just something we figured out on our own. We thought we'd invented the M/S mic technique, too. We thought we'd invented all that stuff!

I'd also help out Paul Davis, a songwriter who had a lot of gear at the studio. He wrote several number one records and he sang on a lot of records. I'd be working with him and he'd say, "Dave, that snare sound is the most amazing snare sound I've ever heard!" And I'd be thinking, "Man, I'm catching on to this engineering thing!" But then he'd say, "But it's just not right for this song." And I'd feel so deflated. There were thousands of moments like that.

So how I learned, you couldn't really call it training. The people I was around didn't sit down with me and say, "Do this" or "Do that." I wasn't really taught; I was inspired. Once I learned what could be done, I could figure out a way to do those things, too. And I spent a lot of time learning how. There was no Internet, no schools; and in Atlanta, although I was blessed to work in the finest studios, overall there weren't that many.

> "If you want to pee with the big dogs you've got to get off the porch!"

Because it was Atlanta, my home, I didn't realize at the time how important all those people were to me. It wasn't L.A. or New York, or London—what I thought of as the big recording centers. It took me years to realize the importance of those relationships. I wish I could go back now and thank all those people. I truly do stand on the shoulders of the giants in the Southern music world. I am what I am in large part because of them.

Another thing I did along the way that taught me a great deal was, I built a studio with Larry. You know how, when a small child is learning math, they run into the multiplication tables? It's pretty much torture to memorize them. But once you do, your life is easier. Building a studio is like that. It's probably one of the greatest things you can do in terms of learning how to become an engineer, but it's also one of the most torturous, horrible things you will ever go through.

My friends who have done it all believe it was a very important part of their process of learning engineering. Many of the guests on the show come from backgrounds that contain as much hands-on building experience as they do other kinds of learning. There's something inherently right about building a studio. I don't recommend it, and you can do without it, but it definitely gives you something you can't get any other way. Just like the torture of memorizing your multiplication tables.

THE CYCLE

It's a funny thing, but during this time it never dawned on me that this wonderful thing called engineering, which I looked at as just part of music, was a career path. I didn't separate music into categories like performing and engineering. I used to design the posters for my band because I thought that was part of the music. Then I did the live sound. It was all just music. I would go do carpentry gigs and it was *still* just music because it allowed me to have the money to buy a guitar.

Back in the day we were all on a first-name basis with the pawnshop owners. We'd go to the pawnshop and it would be:

> **OWNER.** Hey, Dave, how's your dog, how's your girlfriend, you goin' back out on the road?
>
> **DAVE.** Yeah, I need my Les Paul. How much do I owe ya, $150? Can I give you $125 and get you the rest of it when I get my first check?
>
> **OWNER.** Sure, man, no problem!

At the beginning of every cycle around the South—prior to which you'd been broke for a long time—you'd spend money like crazy for the first two or three weeks, figuring you had 8, 12 weeks left after that to start saving up for the dry spell you knew was coming after the tour. But then, inevitably, your agent would call and say, "Uh, these four gigs got canceled." So back you'd go to the pawnshop and hand them the guitar. The guys in the pawnshop actually played the guitars more than I did! It was a cycle, but somehow it never really seemed bad. It was music and it was freedom. It was what we wanted to do.

In our world back then, particularly in the South, we made poverty into an art form. We had fun. We found ways to do lots of things. Everybody around me was equally economically challenged, so I wasn't constantly bombarded by the fact that I had no money. Now, make no mistake, we weren't poor, we were broke. Our poverty didn't come packaged with hopelessness and generations of hopelessness. At any moment in time we could get off our lazy asses and go do construction work—which we did, frequently.

> "You can't eat dessert and cake all the time. Sometimes you have to eat your vegetables"

That was the cycle, that was my life, and it was fun. But at some point, much like it must be for a professional athlete who's been playing high school ball, you want to test yourself against the big guys. You want more.

I had a great lifestyle in Atlanta, but I wanted to test myself, and I wanted to get better. I wanted to play in the major leagues. The impetus to leave Atlanta was primarily musical. It wasn't economic, because if I'd made any money at all that would have been a step up. I just decided I wanted to be in a major market.

How did I decide where to go? Well, I don't like snow, so that ruled out New York. I don't like rain, so that ruled out London, although I did spend some time there checking it out. Nashville at the time was pretty much strictly country and somehow didn't seem like an option, although I respect the musicianship there and everything they do. Really, the bottom line was that in 1989 the place to be for music was L.A.

BROMANCE

DAVE: I flew out to L.A. to help a friend get some songs placed. It was October of 1989, and it happened to be the day of the big earthquake in San Francisco. That was actually of import, because the earthquake made me think, golly, do I really want to move there? It gave me a bad image of what California was like, and it definitely made Atlanta look more attractive. I just stayed a few days and went back home. But I still knew I had to leave. So I went back to L.A. around December, with no money, of course, and I scrambled around just couch surfing for a while.

Back in Atlanta, having had such love for everything music, I would go to studios and get the old *Billboard* magazines before they threw them out. For five years I was reading the three-month-old issues. And like everybody of my generation, I was also a reader of album credits. Reading them and knowing who did what made me feel like I was part of the business. The names I saw listed over and over, like Larrabee Studios and Alan Meyerson, for example, started sinking in. So I was in L.A., I had no work, and I didn't know anybody. But instead of going to the tourist attractions, I went to studios. Those were my landmarks, the things I wanted to see. And I did my best to sneak into all of the big studios that I had read about.

Being an engineer, I knew how to get in. I'd say, "I'm here for the session," and the receptionist would say, "The so-and-so session?" I'd answer, "Yeah—that session!" and go on in. I just looked like I belonged there. And that, actually, is how I met Herb.

HERB: I remember. We were stuck out in the lobby at Skip Saylor Recording. I was trying to get into the business, trying to start my career. I'd met the Whispers, who liked me and asked me to come visit them at the studio. But they were particular about their session that day, so I ended up sitting in the lobby and hanging in the lounge, where Dave and I started talking. Somehow, we connected right away. Although to look at us it's hard to believe, we came from a similar place. We both read and researched. We didn't have money, so we had to sneak-read. I used to do it at newsstands. I would pick big magazines, usually porno, because *Billboard* would fit in them, and then I'd slip out my notepad and write down the names I saw in *Billboard*. For Dave it was studios; for me, it was executives.

So we met, there was a connection, and it stuck. Dave was a white guy who was into black music. He'd worked with James Brown, and he had a friend in Cameo! I was a black guy who was into Black Oak Arkansas; Earth, Wind & Fire; and Hall & Oates. A lot of people didn't cross those barriers at that time. So Dave, like he does, came fully authentic. He was cool, and we had a similar ethos. He felt the same way about excellence that I did.

DAVE: At that point in time, I thought Herb was the most important person I'd ever sat down and talked to. So I was enamored. He may have been just starting out, but he seemed like Irving Azoff to me. I told him the truth: that I didn't actually record the albums; I did the demos—although I had actually produced James Brown on one song.

HERB: Which was also interesting. Dave was a young guy who wanted to be in the game. He wasn't a well-known guitar player, someone like Steve Cropper from Memphis who'd played on a zillion records. He was a young white guy who was doing the grunt work. Doing demos and hanging and doing the gritty stuff. It was just so unusual because Dave was so very comfortable in his own skin. The indelible impression was that whatever it is he does, this is my kind of guy.

DAVE: Herb was like me in that we didn't delineate or distinguish the various parts of making a record. We went in at 10 in the morning and we came out with a record. It wasn't like pro football, where you had a quarterback and a lineman. In the studio I played every position. And in Herb's way he did, too.

Another thing that initially drew us together on a subconscious level was that neither of us distinguished between genres. How we determined what we liked musically was very simple: here is the category of stuff I like; here's the category of stuff I don't really feel. Not dislike, just don't feel. And stuff I liked mirrored what he liked. We liked rock, we liked quirky things, we loved alternative music. We liked risk takers and we liked great singers, like Brian McKnight.

We liked gifted musicians who were out of the mainstream, like Amp Fiddler. I think much of the foundation of our bond was that we didn't actually dislike anything. We just divided music up into stuff we liked and the other stuff. Some things just didn't strike us on an emotional level like other stuff did. To this day we are still the same way.

HERB: In a funny way, our geographical backgrounds and experiences were not dissimilar, and our parents were not that dissimilar, either. On the one hand was this Spanish-from-Florida-to-Georgia-to-L.A. guy and on the other hand this Canadian, Kentuckian, civil rights, Jim-Dandy-to-the-rescue, sing-in-the-Baptist-choir-to-get-to-L.A. kind of person. So we both had this odd diversity—although at the time we didn't even know that word. It's just how we grew up. And it was cool to talk to someone who understood. Because when I was running around trying to get into the urban music business, it was mostly very closeting to me. I was culturally a fish out of water. The first company I worked for had a black CEO, and at one point he said to me, "Herb, don't involve yourself in black things. I think you should just talk to white people."

We were both literally living our business. I think most great relationships have an intuitive connection. The connection between Dave and myself hasn't wavered in 30 years. We weren't always working together. He had different management at times and I went different places. But the minute we picked up the phone we were right back.

Being around Dave just invited excellence. As a budding young manager trying to make my bones, having him on my roster—well, he was a budding young engineer and it took off really fast, with no machinery involved. He just had a hit. It was distinctive. And he got hot so fast. There was a hit single; then his remix took it to a whole other place. The floodgates opened, the phones started ringing, he got hot as fire. And it lasted 25 years.

NICENESS MATTERS

DAVE: Not only did my career take off without me trying, it skyrocketed on a song that had no credits! But what had happened was, when I started out and was canvassing all these various record companies, because I was polite and respectful to all the secretaries, they would try to find me little moments here and there to talk to their bosses. I universally got rejected, but later, when the bosses were trying to find out who had done this massive record, the secretaries remembered that I had been in their office and they were comfortable calling me. Suddenly I was the darling of L.A. engineers. I wouldn't call it opening the door, but I had "niced" my way into a lot of offices. So when the boss said, "Get me the guy who did this," I'd get a call saying, "Dave, my boss is looking for you." It was the version of going viral back then. There was no Internet, but there was secretary net.

HERB: Dave has a certain kind of personality imprint and it has been definitive throughout his career. It's been interesting to watch what he's learned to do with it, and how it's fit into the business as the business has changed. I think in an unconscious way we've both used "nice" as a part of our personalities to extend our brands.

It's not that we're recessive. But we've learned that being nice can be impactful. People like Dave and like working with him. And as a manager, I've had many clients who, if they weren't liked, it actually helped my career. Because I *was* liked. People would say, "That guy's a dick, but Herb's really cool. Talk to Herb."

DAVE: It's also interesting that the first project that blew up was a diverse one. Bell Biv DeVoe, Dave Pensado, Wolf & Epic—black, Jewish, and Southern white. It could not be odder. You could not find a weirder mix of people. And to start with, the referral query came from a guy at the label that had nothing to do with the acts and the producers! It was just all this kind of stuff from Jesus. There again, I was just trying to make something come through those speakers that I thought would entertain a diverse bunch of people.

HERB: The show has been helped by that niceness, too. In our early days we were demo-ing out how things worked in L.A. It was, "How do we use whatever we have?" Some of our vibe is undeniably Southern. Like his Florida/Georgia thing and my weird Canada-to-Kentucky thing. In Kentucky you learn certain things about civility and how you have to interact with people. More important, I think Dave and I don't really judge people. I'm a firm one for saying, "I like that." Or I don't. But I'll take a pissed-off, cussing, spitting Tea Partier and sit with him all day long. People pick that up, and, truth be told, it's turned into a career. L.A. was really like a petri dish where you could develop your personality and really experiment. It still is.

DAVE: One of the things I was brought up to believe is that if someone pays you a dollar you do two dollars' worth of work.

HERB: Can I double my commission?

DAVE: You haven't yet? Back when I was going around trying to get my name known by all the secretaries, my pitch was, "Hire me, I'll make your life easy." I'll do the record, I'll get the files, I'll supervise mastering. I'll come do your yard work, and, by the way, here are some flowers. Not for ulterior motives, but because that's the way we were raised. In other words, to this day if someone hires me to do something it is rare that I don't do way more than what I'm paid to do.

HERB: To that point, as Dave's manager, I tend to tell people that, whatever their timeline, they should throw it out. Because Dave is going to get your project to perfection. That's just what he does. He's not going to let it leave his shop until it reaches the level of his bar. Which usually pleases everybody. But I kind of have to let them know that it's not just that he's being slow. He's just not going to give it back to you until he thinks it's banging.

And me, for example, whenever I'm working with somebody in the service business I do things like get their name off of their name tag so I can speak to them personally. "Hey, Danny, how is your day going?" People appreciate being appreciated. The point of this is, our glue is partially based on being this way. It certainly helped us in the beginning. We both got to L.A. about the same time, our careers took off about the same time, and because of that we just connected the dots. But we really were just starting. He took off, and I had a little bit of heat because he was hot. Then Brian McKnight got hot, and all of a sudden we were dealing with very hot, and it was, "Oh shit. We've got to manage this."

MAKING DECISIONS

DAVE: Life doesn't come packaged with labels. There's not a label maker you can get that will tell you, this decision was bad, that decision was good. Life doesn't come that way; it just comes with the need to make decisions. If you're not making decisions, you are not really going to get anywhere in life.

> "It's okay for you to want to go someplace, even if you're not exactly sure where that is."

At one point I made a decision to go to England. The '80s music coming from there had captured my imagination. I especially idolized the British producer Trevor Horn; I loved everything he had produced. My plan was to walk into his studio Sarm West and work with him. That didn't happen, and I ended up running out of money. That wasn't such a good decision. But I didn't look at it as a bad decision. I went back to Atlanta convinced that I needed to go to L.A. Turned out that was a good decision. Several management situations turned out to not be the best decision. Then I decided to go with Herb, and that turned out to be a good decision.

But Herb and I tend not to label a bad decision a bad decision, because that decision provided us with enough information to make the next decision better. I learned that from him. It's better to make a decision than to do nothing. If you make a decision that produces results, stick with that. If you make a decision that doesn't produce results, punt and make another decision. Decisions are just tools. Failure, or a bad decision, is just a teaching opportunity where you learn not to do it again.

When you go to Vegas and play the slot machines, you don't expect to win every pull of the machine; you expect to leave with money. That's the way decisions are. Don't expect every one to be right; don't expect every one to be wrong. Just expect to leave the machine with money in your pocket. That's life, real simple.

Things may be easier to discern in hindsight. But if you begin to trust your intuition early and it works, then you ride the down and the up. We've actually always been that way, and we've always been there for each other—down and up. Inevitably, we trusted that it was important for us to get through it. Herb's life hasn't been all up; mine hasn't, either. Nobody's is. But we couldn't have told you back then what decisions were the right ones.

HERB: You're just feeling your way through, and then it develops into a pattern you can trust. You don't have anything else to go on—particularly in this business. Because there are different rules, and they change per category. For example, if you work in the pop space it's one thing. But I know people in the touring business who don't understand that on the urban side of the touring business you can die over things. I have friends who are dead because of some petty backstage dispute. That doesn't happen in the pop world.

So there are all these kinds of rules you are divining as you go through. Some of them are really serious, and none of them do you expect. The idea is that your personality can grab hold of things and sort of force its way through. Then people notice and start to ask you, "How do you do it?" Because you've begun to realize what works and what doesn't. It can often be a great leap of faith, and it really helps to have friends who understand. Otherwise you're just out there by yourself.

Sometimes, though, it turns out that you do need to be out there by yourself. For *Pensado's Place*, I had to cut off contact with my friends for two years, because they all, unequivocally, would have advised me not do do it. "WTF? You're not even in the engineering space and you're going to create an Internet show about it and make a living?" A lot of my friends are sort of killer one percenters. I went away from them for two years and when they saw me later, they were like, "Where you been?" I started telling them about what I'd been doing, and they were stunned.

Hopefully as a child you learn to listen to your gut feelings, and your parents encourage and allow you to make decisions. To be a musician in Atlanta, to work on those kinds of records, to go from Canada to Kentucky, and to say, "I'm going to try this" . . . it's risky, and you've got to have some cojones because you'll often be doing it alone.

I had a decision to make when I had a job offer on the table to be a news reporter for a TV station in Lexington, Kentucky. Should I take it, or go hang tobacco to raise money to drive to L.A.? I hung tobacco. Nobody would advise you to do that. There's no parent, no aunt who's going to say, "Yeah, turn down that newspaper job and go run out to L.A."

But when I look back I think I know part of the answer. When Dave and I came to L.A., we didn't come here just to get rich. We didn't even come here to make a living! We came here pursuing something nebulous that made us want to be around music. Not just any music. We wanted to be around great music. And we didn't just want to observe it; we wanted to contribute to making it. I'm convinced that the overall attraction was as much to greatness as it was to music. We just combined those two things.

DAVE: There's a saying, "God loves drunks and puppy dogs." I think one of the reasons he does is that they are at the mercy of the cosmos. Somebody's got to look after them! When you are so pure of spirit and you just come out and pursue something that you have in you, well, that makes you unique. There is a lot of hurt, a lot of pain, and a lot of loneliness when you pursue a dream. But the pursuit of that dream also makes you so vulnerable that sometimes something in the cosmos—let's call it God—looks out after you.

L.A., for me, was a place for dreams, and I'd been dreaming all my life. Girlfriends have left me because I was a dreamer! Back then, it felt to me like I would never achieve if I didn't go to where the best dreamers were. I had to go where I could rub my dreams up against other dreams. That's the only thing I had faith in. I didn't know if I would make money or make a living, but it turns out my dreams found the right place to do that. And the environment was wide open.

HERB: I think it still is wide open. How you execute your dreams has to do with how well you understand the playing field. My connective tissue with Dave was made up of both our aspirations and our value systems. Good stuff like excellence and knowledge and civility bonded us, but so did the bad stuff: hardship, money woes, rejection, career lulls—all of it.

Show business has a lethal get-your-ass-kicked quality to it that few escape on the journey. You either beat it or fold your tent—it gets pretty abject. You dig down and find another reason to put another foot forward. Both Dave and I have been can-of-beans broke—and we've also been millionaires.

Role models were big for us. They gave us hope—a measuring stick and a bit of a road map. Invariably you go your own way and win or lose on your own merits, but Dave and I had the great luck and pleasure of meeting our role models. It shaped us and made us more useful to ourselves—and to others. It even made us role models for some others, which carried great responsibility and still does. In our early years, especially, it was great to have someone you could talk to. Because you can go a little bat-shit occasionally as things happen, no matter how much you think you have it together. An interesting thing is that, in our day, those role models tended to be in the entertainment industry. That's not necessarily the case in today's world. Today your inspiration can come from many different places.

Now you really have to innovate to stand out. It's a crowded global field with tons of brilliant, aggressive people figuring out the next big thing. Breaking though is ridiculously tough. So is just surviving. Part of why Dave and I are so energized is that we are living the experience of innovating and breaking through with all these new tools. Obviously, at our age this is not typical! But repurposing our tons of experience makes us both newbies and vets at the same time.

Now we are playing in three arenas: entertainment, education, and technology. Most valuable to us is that it confirms the importance of dreaming, then executing while following your own North Star, keeping and setting your own bar. Difficult, sure. Hard as hell. Worth it, undoubtedly. Winning with a friend—priceless!

DAVE: If the reader takes anything from what Herb and I have discussed here, it should be that no matter what your passion in life, or what the thing is that affects you in your gut and moves you, you are probably going to have to make a major move, either philosophical or physical, to get yourself in the position to do what you want to do.

The important thing is, don't make that move for economic gain. Make that move for fulfillment. Once you have value to give the world, the money will chase you down. If you go for money, it never works out quite like you expected, and you often end up making counterproductive decisions.

But if you go in believing this is something you really, really want to do, and you provide the people that have the money to pay you with value, they'll hire you and give you some of that money. But you've got to go where they are, and where the opportunity is. That's what Herb and I both did. It wasn't about "I want to go out there and be famous and have number one records." We just wanted to come out here and be a part of something that gave us satisfaction on so many levels. And when we acquired skills that had value to people, they paid us for it.

Success means one thing to Herb and another to me, which is why we are such a good team. Herb respects my need for philosophy and teaching and keeping everything that we do free, but he's a business guy. His responsibility is to find a way for us to pay for all this stuff and to generate things that people want to pay for. Without Herb, I wouldn't be successful. It's a very symbiotic relationship and it's worked great for 25 years because we each know our lane.

HERB: We're into healthy competition: We were organic about what we wanted to do, but we also wanted to win. We saw competition as being a healthy thing. If you didn't do well, you wanted to improve.

You don't always have to be clear about where you are going. It's okay to just want to go someplace. There can be a time of sharpening your sword so that when you compete, you can win. That differentiates those who win from those who don't, because some of it is just desire—particularly for those who don't have a budget yet. Sometimes you just have to grit it out with what you have.

3 HardDrive

The nickname that became
a blessing and a curse.

HERB: The irony, of course, is that the nickname was never about computers. HardDrive meant work ethic, the very same work ethic that not that not long ago almost killed Dave. That's the curse part.

DAVE: I adopted the name willingly, because nobody seemed to be able to pronounce or remember Pensado. It was great that instead people could ask, "How do you get in touch with that HardDrive guy?" The manager I had after Herb thought it was a good branding tool, so he always insisted that they add it to the credits. But you'll notice that once I went back to Herb for management I dropped the HardDrive.

How it came about was, I was doing pre-production on one of the Bell Biv DeVoe re-mixes with Wolf & Epic at Richard Wolf's house. We weren't yet using computers for recording, but we stored all of our samples on a hard drive. We were heading for another studio and I said, "I've got the hard drive." I guess it was my accent, but Richard started teasing me about the way I said it, and that was that. It became a joke.

Over the course of a month or two it morphed into my name because of two things. One of the gentlemen I met through Herb who was very instrumental in helping me out was Louil Silas. Louil would come by the studio at 10 in the morning and he would expect me to start the session at noon. I ended up working noon to noon (around the clock!) in order to accomplish what I thought was expected of me. And it would irritate the hell out of me that people would come to the studio to work, stay for only maybe three hours, and then leave.

That didn't cut it for me because I just couldn't face Louil, or Herb, or whomever I was working with at the time, and explain why I hadn't accomplished what I was supposed to have accomplished. So it quickly became a complaint that I was pushing people too hard, driving them too hard.

I was doing that without realizing it, though, because of the unique and fun qualities of the particular profession of being an engineer. I didn't notice that it was having a physical effect on me as well. I always thought that work and godliness were part of the same concept. It's the way I was brought up: the harder you worked, the closer you got to God. I still believe that, by the way.

I also was like that because, back in Atlanta, we'd had a different kind of work concept. Studio time was rare and precious there, so we utilized every second of it. In L.A., a 24-hour lockout was really 12 hours—that's how much time people actually used. So in my early days and nights here, when people weren't working every minute, I always felt like we weren't getting enough done. I felt like we had to go at it a little harder.

It was one of the BBD guys who pointed out to me that I might be pushing it a little more than was necessary, saying, "Let it go, you don't have to drive so hard." It was one particular night when we were supposed to record some vocals on the ballad remix I was doing for them of "When Will I See You Smile Again?" They showed up around 8 and weren't ready to go until 11, and I was like, "C'mon, guys, I'm paying for this time!"

So the name began because Richard was amused by how I said the words. But people around us assumed he called me that because I was pushing too hard. I never straightened them out, because I'm a big believer in whatever happens, don't correct it, just let it go. Unless of course I'm called "racist" or "child molester," or something else bad. Then I'll defend myself. But short of that I don't really care what people say. Well, actually, I do care, but you can't control that. A good thing to remember, by the way, is that, in the Internet world, the people who try to defend themselves just make a mess out of everything, when, in general, if you just let it go, it drops down to the third or fourth page on Google and starts to vaporize.

HERB: It was the first project Dave really did in L.A., and all of a sudden he had this nickname that stuck. It wasn't about promotion or branding. I had nothing to do with it. It was just that everyone who worked with Dave, everybody who hired him, would say, "Damn, he'll work you into the ground!"

One of Dave's trademarks to this day is that he doesn't really mix for the client. He mixes until he's satisfied, which tends to surpass what the client wants. And then they go, "Oh, my gosh, he's got such a perfection level!"

But that perfection level comes with really hard work, and staying until it's perfect. If that was three in the morning, five in the morning, whatever it took, Dave would stay and get it done. It was wonderful, but we didn't realize back in the day that working that way was damaging to us.

It was an amazing, true nickname and it was natural to him. He loved being in his room and he loved being a mixer. It was his salon, a place where you were social, where Dave helped out many people, and where there were all sorts of discussions about Dave's philosophical constructs. There was also a powerful work ethic on display. Until the incident happened. And then we realized HardDrive was also a curse. Up until then, we hadn't realized the yin and the yang. You can seek perfection in your work, but you can't always be about work. Now we know you have to be healthy and pace yourself. Back then it was just hit, hit, hit, hard. Every day and night.

DAVE: A bit more intelligence could have been applied to the process. We all have built into us the fact that we don't accept our mortality. So you work based on the fact that you are immortal until something comes along in your life that proves you are not. Then you modify what you are doing. I don't work less now. I just work smarter.

THE LAST INTERSECTION BEFORE COMMERCE

DAVE: My job is sometimes almost impossible to accomplish, because, being the engineer and the lowest person on the food chain I answer to everyone: the artist, the producer, the A&R people, the record company president, and more. I have to listen to each one of those entities in the record-making process, but they each have different goals and needs. For example, one might have to do something quickly and inexpensively. Another might have a need to maximize the creativity that's applied to the process. I'm constantly having to inject all of that into what I'm hired to do, which is to be creative, and to take all of this and repackage it so it will appeal to the listener and make it something the listener wants to own in some way. At the same time I have to do it efficiently and quickly. I had to learn all this, of course. Initially I thought you just went in and made a record.

But it's not that simple. Some artists wanted to come in and sing vocals for just two or three hours. That would make me livid, because I'd have the record company saying I had two days to finish—and I'd also have Wolf & Epic, who have two different approaches themselves; that's what made them unique. So I was refereeing disputes between the two of them, and also between the unit that was them and the record company. I also had to referee disputes between the A&R person and the president of the company. Then every once in a while promotion people would come by, and they'd have a whole different set of requirements. So I became known as someone who pushed everybody harder, including myself. That was my solution, and it seemed to work. We got a lot done with that process at the time.

THE SALON

HERB: Everybody came by: label, managers, artists, agents, promotion people. And Dave, with his salon, was just like a maestro. He'd play us all, and then everybody would leave. One way he'd play us was that he wouldn't have us all there at the same time. Each one of us would leave satisfied, thinking, "Yeah, I won. I got my way." The next day he'd make sure the record company was cool and got its say. The next day, the artist was cool.

DAVE: I looked at it as having an opportunity to see what everyone wanted. I also learned that sometimes artists are easier to work with when you can talk to them without their entourage so they can be themselves. The same with managers. Sometimes they can manage more efficiently when the artist is not around. The record company people can explain themselves better that way, too. I realized that I often got better information when I isolated the elements than when they were all together. When everyone is in the room, there can sometimes be competitiveness. It can be difficult to be truly honest in front of so many competing interests.

HERB: If I as the manager of a multi-platinum act was there, and the president of the label was there, well, we both have a lot of weight. The label president may have a very different perspective from mine. But I think I know my act better. We would have to talk about it, and it was better for Dave to hear that unexpurgated opinion. It was like going to confession—you just sat there with Father Dave and talked it all out.

Now, this may sound a little metrosexual, but most people who went to Dave's salon left feeling good. They got all this out emotionally, and when they left they felt like their record was in good hands.

Often Dave was also doing additional production on the recording. He'd replace sounds and make suggestions. So clients had a deep engagement with Dave that led to this kind of longstanding love—and brand—so that even among his peers he had a special niche. It came out through the work, the handling, the care he took with everything. His personal standard was, and is, that you got way more than your money was worth. But in hindsight he paid a price.

DAVE: I'd like to make one amplification: At the end of the day, the various entities that are required to make a successful record all basically have the same goal. That goal can be reduced down to the fact that they want to create something that millions of people are going to adore enough to want to listen to and own, that also represents their creative philosophies and input, that was done efficiently, quickly, and cost-effectively, and that when everyone involved with it hears it they are proud and happy. That's what everyone wants.

Ultimately all the entities had the same goal. They just had charge of various components of the goal. It was my job to synthesize that and put it into the record. And I learned that the easiest way to do that was just to make a great record. Then everybody was happy. So sometimes I had to move them gently and respectfully out of my way so I could make a great record, which was what they wanted.

It's like being a parent and making your children eat their vegetables.At the end of the day we were all happy, not so much because of specific things that I had done in the mix, but because I was able to pull all of the elements together to realize the song's potential. Which, ultimately, managed to fulfill everyone's expectations. But I had to get them all on the same page in order for them to know that.

HERB: Part of the HardDrive mystique was that, because he was so good at what he did, we'd just say, "Do it, Dave," and go away. Then he would take all these competing interests to another level and make a great record. He would do it until dawn or until noon the next day, and for 48 hours and all through the weekend. But we all felt good saying, "Dave will handle it."

THE HARDDRIVE ETHOS

DAVE: I learned very early on, even before I came to L.A., that no one ever hired me again because I did something cheap or fast. That doesn't happen in my profession. The triangle has cheap at the top, fast on one corner, and good on the other. Pick two. That's pretty much what you have to do.

It could be argued, of course, that today's model is completely different in terms of that philosophy. And I think a young cat coming along, no matter what part of the industry he or she is in, has to modify that. But still, what we are trying to do is take the artist's vision and then add to that the producer's vision, and create something that millions of people treasure and want to own.

In a streaming world, "own" has a new connotation, but essentially it is the same: If nobody wants to hear or purchase what we end up with, I've failed everyone. If someone does want to own or listen to it, I've been instrumental in helping shape it and I've done my job.

To clarify, a hit record is not just what I did, of course. That's what I'm saying here. A massive, talented team has to be involved, from the writer to the promotion guy to the program director and the record company. Even when it's a really good record, it's not a hit until the record company fulfills its duties and obligations, and the artist does enough touring, and the promotion person gets enough magazine covers. Even then it may not be a hit. It may be a timing issue with that particular song.

One of the things that keeps us all so engaged and interested in this profession is that it's not like the 100-meter dash, where you can run it and measure it and either you're Usain Bolt or you're not. We've been involved with incredibly great records that didn't sell at all, and we've been involved with records that we thought were just pretty good that went through the roof. What keeps you motivated is, you can't really master this process. Hence the HardDrive method: you just have to bring it all the time.

HERB: Reflecting on our journey, that is one of the points where we intersect: bringing it fully. Whether we're watching a Laker game, listening to a piece of music, or even talking about race—if we work on something together, that's what we do. The way we approach the show is the same. We've gotten smarter about the pacing and handling ourselves, but most of our results still come based on that HardDrive ethos.

DAVE: Could someone who didn't like the taste of ground beef make the world's best hamburger? I'd say about 99 percent of the most successful people in this industry love the records they make. They love the music they're creating, and that passion can extend to every genre. Because what makes music great is not the genre; it's the passion, the emotion, the feeling, the vibe, all of that. Something you'll notice about successful cooks is that they love the food that they're cooking. They know just how to max out every flavor, every nuance, and the temperature at which they should serve it . . . uh-oh. I think I probably just painted myself into a really horrible metaphor.

But the advantage I had being the low guy on the food chain is that I was making things happen around me without drawing a lot of attention to how I was doing that. Successful engineers have a demeanor that includes believability. People just innately trust them. And if you use that trust respectfully, and you prove yourself a couple of times, then the process becomes very easy. Because there is no smoke or mirrors. There is no trickery. It's about convincing the various entities that their goal is my goal—which it truly is. The desired end result is a great record that we can all be happy about and proud of, that will meet the commercial needs.

TRUST IS A TWO-WAY STREET

HERB: When you got involved with Dave as a human, and you saw the work, it was easy to extend the trust. We all know that the way Dave manages us works. From the beginning, there were no shrinking violets. Wolf & Epic, me, Ron Fair—we are all people with strong opinions. So the first thing he had to do beyond the record was to get very strong-willed, already successful people to trust him.

The salon dimension was that, for the two hours you were there, you spent maybe 20 minutes talking about the record. The rest of it was philosophy and theory: Carl Jung, the rule of thirds, climate change, racial matters. Dave is a Southerner who's had a lot of success and has worked with a lot of black people. He has a healthy respect for, and a lot of curiosity about, the culture. So there would often be some very wide-ranging discussions. You'd come out and go, "Damn. I think I was just Daveified. But . . . I feel smarter!" Which is not generally the takeaway from most mixing sessions.

Since the records came out good, and it worked with the big wheels, the little guys coming up who got to work with Dave tended to have a sense of confidence about the process. Very quickly Dave had an awe factor. It has lasted 30 years, but I think it was forged out of hard steel in those early days when there were clients coming in with expectations. Bell Biv DeVoe weren't shy, Brian McKnight wasn't shy. I wasn't, either!

Dave was just this unique person, different from most others in that space. That first record was the end result of a lot of unusual relationships and touch points. That became a trademark of his that has lasted: Dave is expert at taking input from diverse personalities, synthesizing it, and using it. Of course, you wouldn't have that if you didn't also have Dave's need for perfection. Because it was his personal HardDrive that made sure your record was okay, and all the rest of it was a bonus. If somebody didn't have that drive, your record would never be good. You'd get all the fluff—the hype and the reassurances—but your record would be just okay.

DAVE: I just wanted to make great records that matched the quality of records I thought were great before I started engineering. That's all I was trying to do. Herb's alluded to the fact that I loved having lots of people around just to get information. But also, if I didn't see anybody at the studio, I didn't see anybody! I basically had no life. Still don't, unless you drop by the studio.

But an important component to all this is that, in my profession, trust is not a one-way street. Who wouldn't trust Ron Fair's opinion? Who wouldn't trust a 14-year-old Christina Aguilera's opinion? Who wouldn't trust a girlfriend sitting in the back of the room? That girlfriend is probably sitting there because she loves music enough to be with that artist. Trust is not a one-way street. But her opinion is valid.

I always assumed that the opinions I was getting from all of the people around me were right to begin with. It wasn't like I was manipulating information I thought was incorrect. I just assumed that Herb, for example, knew what he was doing. He'd proven himself and I would be a fool to reject that. My job wasn't just to take diverse opinions and my job wasn't to take only opinions I thought were right. I assumed they were all right. That's what both complemented and complicated my job. Everyone can view a situation differently.

Everybody looks at a piece of art on the wall and has a different interpretation of it. All of the opinions were right, but it was my job to find a way to make all of those right things work. So I trusted pretty much everybody who walked into the room, and I found early on that the more people I had in the room, the better the quality of the information I got.

Because, if you can please one person, that may or may not please a million people. But if you can please 10 or 20 people, then you maximize the possibility of pleasing millions of people. It's not rocket science. On the first Christina Aguilera album I felt like she was making a record for her age group. That made me think that some of her opinions carried more weight than those of anybody else in the room. At the time, I had to force some of those opinions out of her; she wasn't yet comfortable in the environment.

I got good information from Christina. She was very, very shy on the first album, but the second album was the Spanish record, and by that time she was comfortable and confident in her opinions, and she asserted them. And by the time we did "Beautiful," she was full-on in charge of her records. That's the way it should be. And she was also smart because she trusted Ron Fair and the people around her who did their jobs really well. And she trusted me.

THE ENGINEER'S CHAIR

DAVE: Now, I have said that the engineer is the low man on the totem pole, but in many ways that's not correct. From the engineer's seat you are controlling a great deal. You are controlling the environment, and you can't show fear. From the lighting to the vibe to what's coming through the speakers and how loud it is, the décor, the CDs you have lying around, everything—in that seat you are in the center of it all and you are controlling it.

When you go to the doctor, he's wearing a white lab coat; he's got a receptionist and the degrees on the wall. If he showed up to carve on you wearing flip-flops, cutoffs, and a wifebeater, you might go, "Whoa! No, no, no. I don't trust you."

What I learned early on, and what I teach my assistants, is that just sitting in that chair gets me trust, much like going into the doctor's office. You don't know if that doctor went to college or not, but you trust him because of the surroundings.

An engineer commandeers the surroundings—but generally in an unobtrusive way—and people are going to inherently trust that engineer unless he or she destroys that trust. Once you start amplifying that concept, and using that little pulpit, it becomes very easy to get people to trust you, because they want to. That is why they are there. They assume, when you are sitting in that chair, that somebody knowledgeable hired you because they knew you were good.

So they want, and need, to trust you, and the commerce is in your hands at that point. Their career, your career, everyone's ability to continue working and feeding their families, is in your hands.

It's important to understand that this intersection, which happens during the mix, where the needs of all of the people come together, is the last intersection before commerce. People often don't really get that. Promotion people meet separately, A&R people meet separately, you make the record independently, but it all converges at that point.

IN THE MIX

DAVE: Often things don't really intersect and come together until the mix, and then they separate again after the mix. During the mix a lot of people have to come together; a lot of people have to agree. And as the competing interests—artists, band members, label, manager, marketing and promotion departments—coincide, if somebody isn't an engineer—in the choo-choo train sense of engineering, a conductor—who makes everybody leave feeling okay, then usually that train will stop. And you'll hear people coming out of the room saying, "I don't like the mix; it doesn't sound right." Maybe they are correct, and the mix is bad. But the train of commerce comes to a stop at that point until somebody gets it fixed.

HERB: You can lose control of the session if people lose confidence that you're going to be able to accomplish what needs to be done. That's a problem. It's also a problem if you find out at that point that the recording isn't what it needs to be. You could have complete control of the session, but if it's just become apparent that the recording is lacking, it takes a very high level of honesty to communicate, "Uh-oh, we'd better go back and fix some things." That can be a very delicate and very challenging process.

> "Mixing is a team sport."

So, often the mix is where things either move forward or stop. When you make a record, many times it doesn't really get evaluated until you come to mix. You're battling along the way, but sometimes you're also thinking, "Let's get a mix on it and see where it is." And at that point in time some mixes reveal, "Whoa, this is not happening. It's not where it needs to be."

DAVE: It's like drawing a piece of clothing versus seeing it on a model.

HERB: Absolutely. It's a whole different thing. I've been there many times when a song has come out of the mix, and you're sitting there with the people who have the responsibility for voting on whether millions of dollars are going to be spent or not, and where that money is going to be directed. And the guy who's making the mix is sometimes asked his opinion: "Do you think this can be fixed?"

The psychological part of this has always been fascinating to me. It actually has a lot to do with your commitment to what is often a very nebulous process. It's got to be in your bones to go that far, to manage all those personalities and make all those decisions. You can't just do it intellectually, for the money. You can try, but you can't get far that way.

I've been privileged and lucky in my life to have close relationships with people in all of the various responsibilities at all of the various record companies. And I can say, without equivocation, that every single one of those entities has a passion for music that's admirable. The concept that record executives are heartless, soulless bean counters who only want to make money is completely false. It's true that they are often tough and driven. But pretty much my experience has been, if they are the ones touching the creative process, everybody literally has the same overall goal.

An interesting aside is, all of these entities once provided a filter system that kept us from being exposed to less than quality music. In today's world, we don't always notice it, but we miss that filter. "Quality" can mean anything to anybody, but what was good about that system, with all its flaws, which were numerous, is that it had a way of keeping most of the crap from getting out. There were multiple redundancies and checks and balances involved, and it is certainly true that sometimes they missed the good stuff. In today's world we have equally good stuff, but we also are exposed to way more crap. It's getting harder to find the good stuff. It's like looking for a needle in a haystack. Anybody can put their video up next to Rihanna, but so what? Since the filters the infrastructure provided in the past have, in many cases, been removed, it's gotten chaotic.

That dust will settle eventually, and when it does we'll find the new model for how we consume music will be the filter that protects us from all the crap, and that will usurp the position of the record company. In a lot of ways I miss that model, but more people have access now and that's a good thing.

WHAT I TEACH ASSISTANTS

DAVE: What I teach assistants is kind of what this whole book is about! I try to teach them early on that, for me, they are being paid to sometimes have an opinion, and sometimes to subvert that opinion. The hard part is knowing when to assert your opinion and when to say, "You know what? I think you're right, that's better." That's the earliest thing I teach my assistants, how to master that art. And again, that's all predicated on the concept of trusting the people around you.

One of the best mindsets to adopt for that process is: sometimes it's quicker to just do it than to discuss it. I've noticed over the years that when someone has an idea that I disagree with, rather than discuss it, it's much quicker to implement that idea. Then we all just listen to it a few times. Probably 90 percent of the time, if it's a bad idea, someone other than me will bring that to the attention of the group. And if it's a good idea, it's usually pretty obvious.

Doing this removes me from having to be that bad guy who tells someone that his or her idea isn't any good, and then having to defend my position. It's quicker to just try the idea. Usually they will hear whether it is good or bad. A beauty of that philosophy is that many times a new idea will be spawned by trying that bad idea that turns out to be better than any of the original ideas.

The ultimate philosophy in taking that position is that you want to provide an environment where creativity can flow and not be hindered, and where people don't have to feel embarrassed. I've sometimes worked with people who tended to be so respectful that they wouldn't voice their true thoughts until I left the room to pee. Then they'd ask my assistant, "Do you think it's okay to tell Dave to bring the snare down a little bit?"

So I have to create an environment, not just for the big shots to feel comfortable, but for the little shots, too. Because great ideas can come from anywhere.

SETTING EXPECTATIONS

HERB: To your point, Dave, about the newer young clients, it's even more important to to make them comfortable, because they are so in awe. They are nervous. And in order for them to feel good about the money they spend, we have to spend time making sure they feel like there's an open door so they don't ever walk away embittered.

About eight months ago we started adding a creative phone call or Skype to the process before we begin. That's changed everything. We talk through all the creative stuff, expectations and deadlines and how we're going to transfer files. By the time we finish that call, they feel comfortable and we're friendly. They've met Uncle Dave, and they enter into the process feeling good and ready to feel good about the consequences because we've set up how it's going to go. That's especially important now, because working with mixes over the Internet is very different in terms of communication.

DAVE: Something I recently learned that is germane here is that because of my enthusiasm for and commitment to music, and also my desire to keep work coming in, I sometimes gave the impression that I was overpromising what I was going to do. What I learned, from Herb, actually, is, don't change what you do, just don't give it all away at the beginning. Show them what you're going to do as you are doing it. I guess the phrase is "under-promise and over-deliver."

That was a hard lesson for me to learn. Because when I first hear a song I'm going to mix, I have all these ideas gushing and I have expectations so high that sometimes I can't meet the expectations I myself have created. Now I just wait for the moment to arrive, then talk about it and do it at the same time. The problem with expectations is that not everyone can afford the time or the money to fulfill a tendency to overachievement on my part. Sometimes I don't have the time I need to do the exact thing I want. We can still get a great mix, but we have to fit it into the time available.

HERB: It's been a process. With Dave, you get philospher and conversationalist and all kinds of added benefits. It is very easy, particularly for newer clients, to get seduced. He's not trying to seduce them, he's just Dave. But today, when we're working over the Internet a lot, we have to look at things a little differently.

With label mixes, there's an A&R person on top of things. They're at the studio, talking to you. They speak your language and you know them. A guy who sends you tracks from Brazil is sometimes just in awe that Dave Pensado is working on it. So Dave could say, "I'm going to take a holiday and go to Pamplona and run with the bulls, then I'll come back and spend a day on your mix." And new clients may tend to say okay. But they never really feel that way. They just want the mix. We realized that in the beginning we were doing too much conversation about the potential.

DAVE: The client base for my profession has changed because previously only a few elite professionals had access to the tools. Nowadays everybody has equal access to all the tools that all of us use. So on Monday I might get a record that was pristinely recorded by some of the greatest engineers to have ever lived, and on Tuesday I get a record that was done in a bedroom that's not the best thing that was ever recorded. The expectations for what leaves my room are identical, so I have to figure out how to resolve that issue within the time frame and budget that I have.

Because I am blessed to be able to make records and then let Herb sort out all the casualties, bodies, rewards, and comments, the way I solve it is, Herb solves it. I work until Herb says, "That's enough. We have to move on."

HERB: And now we solve it before the process with a conversation: here is what it is going to be; here is what you should expect. It may be off 5 or 10 percent, but you have to set a realistic bar. You can't have an ethereal bar. You're still going to do your best work, but the bar is there and everybody is aware. It's the same as when someone tells you about a great movie they saw: "Oh, you're going to love it; it's the best thing ever." Then you see it and it's just okay. Instead, beforehand, you can work together with somebody, set the bar, hit it, and make the whole process much smoother.

ADJUSTING THE HARDDRIVE

DAVE: It's an interesting world now because even the major label projects I've been working on lately are a bit more difficult to get to what I consider the quality that people are paying me to receive.

HERB: Managing expectations is adjusting your HardDrive to what is happening now. The processes have evolved just like the business has evolved. We've found a process that works for us, probably affected a lot by the show. *Pensado's Place* is a natural extension of both Dave and Dave's salon.

DAVE: To take this thread of trust on through, what's happened with the show reminds me of what happened in my early career. What Herb calls Dave 3.0 is that every week I sit across from some of the greatest, most creative audio minds the world has ever seen. And I've chosen to assume that everyone I sit across from knows more than me. I want to know what I can learn from them, and, as it turns out, that's what my audience wants also.

HERB: Which is different from the past. Remember, in the original studio salon, people came to Dave for his expertise. Here, we have Dave creating a road map into the guest's expertise. That was intentional. *Pensado's Place* wasn't designed to be a starring vehicle for Dave. Instead, it was, "Let's take this unique character, this avuncular persona, and make it entertainment." Dave's actual part is only about 20 minutes of interviewing. But we have worked very hard on those 20 minutes. He gets in, asks good questions, and gets out of the way. People tune in to see what Dave will get out of that guest.

Dave is best in situations where he can be unfettered and philosophical, but you can't do that for an hour. It may be interesting to the people who love the particular topic you're expounding upon, but that's a small portion of the audience. We have an audience that is trained to consume and determine within the speed of a screen swipe or a click. So if you're expounding or ranting, you're liable to lose your audience.

At the beginning, we were trying to figure out what to do with this Internet television show that we thought had to stand alone. And what we turned it into is all these disparate but connected elements. Now our producer, Will Thompson, has fans! I have my own fan base and community. We wanted the show to stand on other merits, because it allowed us to experiment. We have replaceable parts, and if we happen to break down in one area, we won't lose the whole thing. When we add new segments like worldwide correspondents, studio tours, or "Recall," it progresses into even less of Dave and me. It's letting the people power the show. We maintain this is addition by subtraction—increasing the value and interest of the show.

I think many people who want to create a show don't look at themselves from a TV audience's point of view. If I sit down to watch TV, there are very few people that I will want hear talk for an hour. You've got to move, you need compelling stories, and you have to attract. Especially with mobile technology, things have to happen fast. From day one we pretty much knew that we wanted it to be that way.

Everyone today is a content creator, and by extension, so is our audience. I tell every person who asks me for advice—which is a lot these days—that they are no longer in a specific subset of a business. They are creating content, utilizing a screen in some shape, form, or fashion. They are likely, even if inadvertently, publishers as well. It doesn't take a lot of deductive logic to ascertain that you should learn how the process works and how to monetize it.

The game has shifted and you can't pretend it hasn't. How you choose to target is, of course, your call. You have the right to compete to see if you can become that two percent that just works on records, but you're limiting your opportunities to win if you don't learn other aspects of the audio profession. Learn it all! The best schools teach that way now, and that's how we pulled this off. You will pull off your thing, whatever it is, differently in the future. What happened for us was, we didn't approach the show like two old dudes looking at the past. We peeled back the covers, then looked down the road.

DAVE: I've noticed that of all the reality shows that one can watch, we're the least scripted. Our show is actually the definition of reality. We are scripted in the sense of the flow, but within that I don't have lines that I have to say or that I've rehearsed. And the most real part of the show is our guests. Each week we have a little mini reality show within our big reality show, because they tell us about the reality of what they had to do to get where they are, and how they do what they do.

HERB: With the show we are an example of a trend we could call adaptive learning. It started in real time with real experiences. You can chart it, and affect it, and learn at your own pace. When Dave does an ITL ("Into the Lair") and posts it, I may be way behind and you may be way down the road. It will be a different experience for each of us.

Khan Academy and Newton, along with a few other smart people in that area, have pointed out to us that there are elements of what we do in the show that refer to that kind of education—adaptive online learning. We have been called the *Charlie Rose* of audio, which I personally love. It's hard to watch Charlie without learning something! Guests are comfortable, so they expound differently. They are more open and giving. Charlie is a genius in making that work for a superwide swath of guests. We attempt the same.

The technical aspects of the learning segments in the show were planned, but what wasn't planned is how people saw them and how they adopted them. That's what's so pleasing—watching how people adopt things in ways that we didn't expect. By the way . . . we have also been called the *Inside the NBA* of audio!

DAVE: I'd like to expand on that last part: the future of the world we live in. It's hard to make money in the new world with an old model. What we are doing is specializing in this world of audio, and there are a lot of people who have an interest in the various forms of audio. If you combine that with our deep delve into creativity, I think we haven't even scratched the surface of our potential audience. That's the beauty of the future.

HERB: We apply digital tools combined with analog analysis to create a hybrid approach, which is partly what differentiates us. The labor involved in doing that, however, is intense. A one-hour show has its own particular dynamics. I have gained enormous respect for those who do it daily or weekly. Production, elements, roll-in packages, location shoots, editing, guesting, research, sponsors, campaigns, staff needs, clearances, emergencies, editing, approvals, platform partners—and then we need to perform on the show! Your metrics and comments are coming at you every day. No hiatus. And this is our side gig! We just try to think it through and keep the business and the momentum going.

DAVE: Over the last 25 years we both have done certain things certain ways because we thought it was the right way to do it. The genius of *Pensado's Place* and the way it's constructed is that it is an extension of our philosophies. We is what we is.

INTO THE LAIR, NUMBER 70: FIVE THINGS THAT INSPIRE ME TO MIX

DAVE: It's helpful to see what our colleagues are doing—everybody needs sources for new ideas. We're all part of the same community—no matter what level we're at—so sometimes it's fun to find out what a guy starting out is doing. It's also fun to see what people I don't know are doing. Inspiration can come from any number of places, and before I start working each day I usually do some of these five things I'm going to share with you.

1. One of the first things I do is go to Beats Music and Spotify. Both of them have their uses. In Beats Music, I have a list I've made that I can just let play top to bottom. Or, if I'm in the mood for country, I'll check that out. There are some songs I use just to get my head right. I also like to hear new releases, and these streaming services give me all of this. I like

the premium versions, because you get a little better fidelity. With both Beats Music and Spotify Premium, you get to stream at 320 kbps.

With Spotify, the first thing I do is go to what's new. I like the *Billboard* Hot 100 section, which gives me an idea of what the world is listening to and liking. I've also got a list my friend Dylan Dresdow sends me of some things he's listening to, and I have a couple of other friends I share stuff with. A lot of times I just let these things play while I'm loading in a new plug-in or waiting (forever!) for my rig to boot up.

2. Another thing I like to do is go to the *Sound On Sound* (www.soundonsound.com) website. They've got a section called "Secrets of the Mix Engineers," which my buddy Paul Tingen writes. I'll pick somebody I'm not that familiar with and learn a little bit about him or her. Like today, I ran into Mumford & Sons. You can find out a bit about where they were at the time they made a record, and they also describe their process, the equipment they used, and some of their settings. For example, they show the Pro Tools session. I just enjoy seeing what other people do. We love this profession and there's great information about it available. You can learn, and also steal some techniques!

3. I also love to go to Beatport (www.beatport.com). In the electronic dance world there are no rules. It's probably the most open musical form there is right now. So let's say I'm about to work on a pop tune and I'm tired of my same old effects. I'll just go to Beatport and let it play. It has popularity ratings for different

songs. I may hear a delay or a weird little sound that inspires something in me. I use Beatport like some of you guys reading this use "Into the Lair." I go there for some new ideas. It's always good to see what the electronic world is doing. Even if you're a rocker, you'll find applicable things.

4. Next—and I know this sounds obvious—there's a lot of usable information on the manufacturer sites. The plug-in manufacturers, especially, are really good about showing you how to use their products and posting tutorials. Waves is incredible, UAD, SoundToys, everybody. For example, I'm on McDSP's site a lot (www.mcdsp.com). I like to go to the artist or the endorser sections. Every site has a user's section, although they may have different names. So let's check out the user stories on McDSP. Here's Crystal Method! I know Ken (Jordan); let's see what they're doing. They talk about how they use a particular set of plug-ins, but the concepts apply to any plug-ins. They also break down a couple of songs they've done recently. Looking at all this makes me feel like part of a community. These are the people who do what I do, so I like to see what they do.

5. Another website I use is www.pro-tools-expert.com, which, if you're into virtual synths, or, of course Pro Tools, is a great site. If you don't use Pro Tools, it may not be perfect for you, but you should still check it out, and then find a similar website for your DAW. Russ Hughes is here on the site, and all his stuff is good. Here they've got "Public Beta" and one of my favorite

things: free plug-ins! They're free for everybody and Russ checks them out so you know they're good.

This helps me keep up with Pro Tools–specific bugs and specific releases, and it also helps me time my purchases. Because if there's a big product release coming out, I may want to sell some of my older gear before the new products debut so I'm ready to buy the new products.

> "Mixing is also just details!"

Technically, this would be Tip Number Six! I'm a guitar player, so I love the website www.guitargearheads.com. I don't get to play guitar as much as I'd like to these days, so it's fun to get to see what's out there, what people are doing, what's available. Like I said, if you're a keyboard player, you may want to go to a keyboard-specific website.

In combination, all of these things get my day going. Some I use for inspiration; some of them I use as a way to keep from getting bored while I'm waiting for a plug-in to load or for a mix to download or upload. It keeps me current and inspired, lets me know what my friends—and people I don't know or people I want to meet—are doing. It just generally keeps me up with this thing we call engineering that we all love so much.

Go ahead and give me a list of some of your inspirations! I take them seriously, because I learn as much from you guys as you learn from me!

4 Breaking the Magician's Code

The secret is not to keep it a secret!

HERB: Traditionally, in order to protect the livelihood of those who practice the art of magic, magicians have only revealed the secrets of their craft to other magicians. The same has sometimes been true of music mixers. But a through-line in this story is that Dave has always been into sharing information. That was the essence of the salon. It's his style of working in the studio, and it's also the essence of *Pensado's Place*.

Right from when he first started having hits, Dave also started getting known for sharing his engineering knowledge with pretty much anyone who asked. From very early on with the Internet, his e-mail address was available and he'd answer questions from anyone who got in touch. Not long after that, a pretty seminal career event for him was an interview in which he gave away what some folks at the time considered insider information. There weren't very many mixers doing that kind of thing then, and not everyone reacted positively to such a public sharing of engineering secrets.

> “Trust is a not a one-way street.”

DAVE: It didn't seem like a big deal to me. I was answering maybe 60 e-mailed questions a month, a couple every day. I enjoyed it. It's difficult to do that now, because of the growth of the Internet and our profile on the Internet. There just isn't time anymore. But the response when I can do it makes everything worthwhile. All the work we put into the show, the long commute to the set, having to work late on Wednesdays after the show—all of those issues become invalid when we read a couple of the e-mails from viewers. You can't read some of them without crying.

HERB: I think from the earliest both of us enjoyed teaching. It's part of our makeup. We just like to do it—there was never any intended effect. Now we hear from folks about what they are up to and they give so much back that it's hard not to get emotional.

I think that ties into some of the psychological stuff we talked about previously. In the mix world, where you handle all of these different personalities, drawing people to you, helping and sharing with them and watching them grow is a consistent line in our professional careers.

DAVE: I agree with you, and by the way, there was a compliment hidden in there somewhere! Like I said previously, back in Atlanta Larry Turner and I started engineering together because we'd both discovered at the same time that we'd taken the band thing as far as we could. And every time Larry and I learned something, we shared it with each other. Then Ed Seay, who'd already had multiple number one records, came into the equation. I'd hang out with Ed and he'd share stuff. Of course, at the time I didn't have anything to share back with someone like him, but I pretended I did. And he was kind enough to entertain me even though he was probably thinking, "Isn't this cute? Dave thinks he discovered that thing we've been doing forever."

Also, most of the studio owners at the time were engineers and it was in their best interest to share knowledge with us. Because the more we learned, the more we worked. The more we worked, the more we brought business to their studios. That was the climate I came up in. When I got to L.A., I found it to be a little different among engineers. Some were into sharing, and some weren't. To me, the ones who weren't seemed unusual. Not that I was a saint or anything. In my makeup, there is a fine line between sharing and bragging. My ego is gratified when I tell somebody something they've never heard before. That's just fun for me

There were definitely people who shared. Alan Meyerson, Craig Burbidge, Jean Marie Horvat, Tony Maserati, and many others helped me a lot. But probably the greatest sources of knowledge at the time in L.A. were the assistant engineers, because they had watched all the engineers work. I could easily ask them, "What did Alan Meyerson do the last time he was here?" Or "What did Jon Gass do? Or Mick Guzauski?" Of course, other people were asking what I was doing, too.

But at some point in time, it seemed like about 95 percent of the assistants shut down. They wouldn't talk. And engineers started setting everything back to zero when they left the control room. Or they'd put their effects on weird presets they'd never use. Assistant engineers were being told that they couldn't share anything. The engineers leaving the room didn't want the engineer coming into the room (namely me!) to know what they were doing. This was especially true for effects, not so much for EQ and compression.

HERB: From a manager's perspective, I'd say that during that time, the label did indeed select engineers for their particular signature, or what they perceived as their particular signature. I sat in on a lot of meetings where people were deciding which engineer to use based on their secret sauce. So naturally the engineers wanted to keep that signature secret.

DAVE: When you think back, it's actually funny that engineers were selected because of some perceived knowledge that they had that everybody else didn't. The truth is, nobody really had any secret sauce. They maybe thought they did, pretended they did, or marketed themselves as if they did, but they didn't!

HERB: But that was the marketplace at the time. Call it secret, technique, signature, whatever modifier or adjective you choose. We were in the process of seeking commercial success. And if somebody felt there was a way to enhance his or her success, it was our job to pay attention to that. Because if you couldn't bring it as the A&R person or the manager, and you didn't know how to say, "I need you to do this specific thing," or "Let's treat the synth pads this specific way," then you had to count on the engineer. So at that particular time, from my chair, engineers had a specificity—a sound, a signature, a something. Who knows what it really was.

"Commit to the process!"

With Dave, the secret sauce was really additional production. He would often go in and change parts in the mix and the artist would say, "I like that better." At any rate, from early on Dave's openness was part of what defined him. But back then, it made some of his engineering peers uncomfortable.

DAVE: I once had a top engineer tell me that I was hurting his income because my rate was too low. Now, at the time I didn't think you could charge as much as he did and look people straight in the face! But you could, and I definitely went up another thousand so I was equal to him.

But, as Herb will tell you, I've never really had a rate. If I liked the song and the people, and I thought there was a good possibility they'd use me for the long term, I'd drop my rate considerably.

But let's look at the concept of sharing, and I wish there were a better word. Because what we are talking about, and what I learned back in Atlanta, is not so much sharing as it is that people who really have a passion for this are so damn excited about it that they discuss it with their friends and peers.

HERB: Which ties us into this chapter. Dave has generally been horrendous about the marketplace. He's a soft touch and cares about people. Somebody can say, "My niece's leg is being amputated, but I have to have that mix before the leg is amputated." And Dave will say, "I'll get it done by tomorrow."

DAVE: Well, if I made a lot of money mixing an album for an artist, and for whatever reason the record didn't do well, and maybe the next record didn't do well, either, then the artist got dropped. Now he's trying to make a comeback, and if he asks me to do a couple of free mixes, I can't say no.

HERB: Dave has his personal way of doing things. His career has always been organic. It has never been masterminded or strategized. He just does great work.

DAVE: I have always had the desire to be the best and to really like the records that I made or was a part of. But I never had a desire to be the wealthiest and most highly compensated guy. I guess that's left over from my hippie days. I just felt, and do to this day, that whatever your line of work, you need to put your heart in it and let your passion come through. Do it the best you can and the money will find you. You can put your rate above everybody else's, and you can make a lot of money if you have hits, but you probably won't be anywhere near as fulfilled and happy as

I am. I love my career and can't wait to come in every day. Not that the money isn't important! It is.

HERB: With Dave it is never about coin. It's always that he's excited about it and it just sort of bubbles over. You'd be surprised if I described the calls that go back and forth about the littlest things, where we revert to being 12-year-olds: "Can you believe the letter we got?" or "Hey, this guest just confirmed!" It's important to have that human side because we are in such a technical space.

DAVE: Probably the question that everyone should ask on the topic of sharing, but nobody ever does, is "Aren't you creating competition for yourself?" For example, Herb in his career has actually created many of his competitors in the management world. But they never turn out to be Herb, because everybody's personality is different. Herb shares everything with them, and has always enjoyed helping people get placed in the positions that they really want. I call them Herb's children and they call him Uncle—or sometimes Obi-Wan Kenobi.

There have been at least 10 executives who got their positions with the help of a phone call or other assistance from Herb. When I came along, I noticed how he handled his assistants and how he brought them along. He had a person who was cleaning his house whom he elevated to a position of power within his company. Sometimes he's been hurt when people he's trained have kind of turned on him a little bit. But they always come back. Eventually they'll turn around and give him credit and thank him and help him. It can be a tricky, somewhat dangerous thing to do. But he does it well.

In my world, it's simpler. I've had quite a few successful assistant engineers over the years. And I think if you piled them all in a room, they would agree that they don't sound like me. They've seen me work; they've got all my settings, my presets, and my philosphies. They've got everything about me that a person could know after thousands of hours with me, and they still don't sound like me.

What that tells you is that if I share a few things over the Internet, I don't have to worry. I actually did an experiment one time. I finished a mix (that, by the way, ended up being a number one record). And I told my assistant at the time that instead of clearing the board he should leave everything the way it was. Then I asked him to redo the mix. Of course he asked why, but I insisted and told him to mix it the way he wanted to. Then I left the room. When I came back, it was completely different from my mix. He started from where I left off and he could have just tweaked some minor things. But his taste and his ego would just not let him do what I had already done. Instead, he was compelled to redo the mix in his own style. And that's why it's okay to share everything you know—first, because someone who copies and uses your information simply cannot sound like you. It is taste that ultimately makes us sound the way we do. Second, if they have any kind of ego at all—and without ego and passion you can't work at the top level—engineers innately have to insert their own taste into the mix. So it's a non-issue.

It's actually kind of frightening how I can pick out the work of my assistants. When I hear their work on the radio, I can tell. I can have a hard time recognizing my own work sometimes. I might hear something sort of obscure on the radio and think, "That sounds great! I wonder who did that."

Then I realize it was me! That's kind of a cool moment. But a lot of times I'll hear something on the radio and I'll guess that it's Jaycen [Joshua] or Dylan [Dresdow], and when I look it up I'm right.

THE ENGINEER AS MAGICIAN

DAVE: As engineers, we are in a profession where nobody really knows how good we are. And I say that humbly. Mostly, only the other engineers can hear the difficulty of what you've done and the skill it took to do it. It's a lot like the challenge a doctor might have. What makes you credible and allows people to place their trust in you? You'll notice that all doctor and dentist offices look alike, because they're trying to show they belong in the profession. When was the last time you went to a doctor's office and asked to see his diplomas and degrees? "Hey, I really need some proof that you truly are a doctor." We get fooled all the time, according to the news!

If they can present a certain look, a certain atmosphere in their offices, you're going to accept that they are who they say they are. They've probably also got some things on the wall that we assume are degrees, but I've never walked up and read one. Nobody does! They could say graduate of Hamburger U. from McDonald's for all we know. There's a reason all doctors wear white lab coats. At that moment in time we assume they are doctors. The office is right, the receptionist is right, the magazines are appropriately out of date—and about things I have no interest in—so I let the doctor work on me.

In the engineering world we have the same challenge, but we solve it in different ways. If you survey the landscape of great engineers, you'll see they usually tend to exude confidence, and they often sort of stage their work area to create a vibe that shows off their unique personalities. This somehow also has a way of making you think they're great! They've got all this gear, and a lot of them still have big consoles and a lot of blinking lights. And there's this hushed quiet. You have to be quieter in a studio than you do in a library; if you talk, you're going to interrupt them. They've got this mad scientist thing. The greats use that and do it really, really well.

There's one of the greats who will not allow anything in his control room that has any woman's picture—name, face, whatever—visible when a female artist is coming to approve the mix. Because he wants her to feel that she's the most important person in the world. Even though he's worked on many hit records with other women—and that may even be the reason she's coming to him to make her record—he doesn't want her to feel that she's competing with anyone. He wants her to know that, at that moment in time, she truly is the most important thing his life. And that's a good thing. So the greats know how to manipulate this uniform we wear.

MAKING ARTISTS COMFORTABLE

DAVE: I think my personality is just kind of designed to be pleasing; I have a need to be liked and to please people. When I'm not liked and don't please people it bothers me for maybe a couple of hours, and then I don't care. But when I'm with them, there is an immediate need for gratification that seems like a constant in my life. The more instant the gratification, the better. I'm sure that's why I played in bands all those years; I could play something and immediately I got feedback from an audience. I like having people around. When people come by I can get instant feedback, and that helps me take more chances. If I'm on my own I may stay within certain limitations since I can't look at the artist and say, "Is this too whacked out? Is that too crazy?"

Sometimes the artist will say, "No, it's great, do it! In fact, do this also . . ." And that idea maybe is an even better idea than mine. I guess you could say my technique is to make them feel part of the process. I encourage them to contribute. Some engineers are a little uncomfortable with that—maybe because they're afraid they'll be asked to do something they can't. My experience is that you can pretty much do anything. But you may put yourself in a little bit of an uncomfortable position, like maybe having to say, "That's the worst idea I've ever heard."

"It's usually quicker to do it than to discuss it."

So, as I've said, "It's quicker to do it than to discuss it." If somebody suggests something crazy, just do it. They'll hear it, and maybe they'll think it wasn't such a good idea after all. And I don't have to say anything. To me, the process is not so much about figuring out ways to impress people—or to make them feel like they are in a real doctor's office—it's just that I love the job. Most of the people I know who do it feel the same. They just love it. And that permeates the atmosphere they work in. Everybody's personality causes each of them to be reacted to differently. My job is to take that natural reaction and utilize it within the context and framework of what I'm trying to accomplish. Which is, I'm trying to take what the artist and producer have spent a lot of time and effort creating and to understand the vision they wanted, then to finish the execution of that vision. That's what I do. Sometimes it's easier to finish it alone. Sometimes it's easier to finish it with a roomful of people. Every mix is different.

HERB: Dave's personality and style of working with people has always been conducive to collaboration and openness.

DAVE: Thanks to my mother, I have a strong mathematical background. I look at it as pure and simple math. If I make one person happy with what I've done, and then I have to present it to the masses and a million people like it, that's cool. But if I make two people happy and I present it to the masses, I've cut my odds of a million people liking it in half. If I make four people happy, now I'm four times more likely to be successful in the marketplace. If you are one guy working alone you've got to just pray that people are going to like it, and you are dependent 100 percent on your own taste translating to the marketplace to produce results. I'm comfortable doing that, and I've got the successes to prove I can do it, but success is not just about sounds. It's about making records come alive and realizing the vision of the artist. If the environment that you create makes people feel good, it helps communication between people. Another, more succinct way to say this, although I'm exaggerating a bit to make a point, is, don't ever work with people whose talent you don't trust. And when you go to mix, trust their talent. It's pretty much that simple.

Hey, I've trusted random visitors who happened to be in the control room to give me an opinion. A very important song in my career and life was "All My Life" by K-Ci & JoJo with Roy Bennett, a dear friend of mine who has passed and who I miss daily. We were working and Roy's girlfriend was in the room and she said, "I'm kind of digging that knock, knock, boomy thing." And I thought to myself, "It's a ballad. Should I really turn up the kick drum?" Then Roy jumped out of his seat and said, "Yeah, man, that's great." So we ended up with probably the loudest kick drum on a ballad ever up to that time. To this day I feel in my heart it was because Roy's girlfriend wanted to hear that knock, knock, boomy thing a little louder.

HERB: The point being, if you have a collaborative nature and you stay open to things, all of a sudden there are rays of light coming through the place. And not only is that good for the people who have to make decisions, but once people know that you are open, your working environment becomes much more positive. That's always been a focus with Dave. It's important to our relationship and the way we operate. If people know that you are collaborative, they come into your room differently. They feel comfortable and you end up gleaning information that makes you better.

DAVE: I'm going to go back to a couple of things that are very important to me. I think you could easily divide all mix engineers into two categories: those that you hire for *their* sound, and those that you hire for *your* sound. In the early '90s, you hired engineers for their sound. I wasn't totally comfortable with that. I already had number one records, but I still wasn't comfortable with "my" sound. My sound tended to be derived from the people I worked with—like when we had those first number one records. They defined the sound; I executed that vision. I liked that, so I tried, and still do try, to make my career about being the guy you hire to help finish your vision, to help create your sound.

One producer I thoroughly enjoyed working with was Polow Da Don. Because I met him pretty early on in his career, Polow and I had the luxury of spending a lot of time in the mix process—and thank goodness Interscope paid for it—trying to get his sound defined. Those are some of the most fun sessions I've ever done, and they also produced a few number one records.

> "I haven't had this much fun since the pigs ate my brother!"

So, if your goal is to try to help people get their sound, you pretty much need to listen to them and collaborate with them. If you want them to just take your sound, then you probably don't want them around, because they're paying you for your sound and you don't need them there.

WHO YOU ARE IS PART OF THE PROCESS

DAVE: I think I am the way I am partly because I came from working in bands, where I couldn't go and perform in front of people alone. I needed band members. Collaborative effort was my world for the first half of my life. And in the South, as I've said, they aren't really forgiving if they don't like your band. We needed each other just to get out alive sometimes, and we always needed each other to entertain the crowd.

A lot of these formative things crept into my engineering. I think that's the same for everyone; we are both victims and beneficiaries of everything that's happened in our lives. Mixing records is a profession where that is also true. No matter what your education is, it's going to contribute. No matter what your personality is, it's going to contribute. No matter what your personal level of wrongheadedness, it's going to contribute. So is your psychological ability. I know guys who are really good at picking up girls; they use the same techniques in the control room with their clients. Everything works. You use your entire life in mixing. I think that's what makes it so much fun. Herb's profession is like that, too.

HERB: It's just natural that you are going to bring your life to the creative process. The people who break through and do extraordinary things are the ones who learn to grow without losing who they are as people. People who lose who they are often wash out. It's tough to hold on to who you are, especially in the beginning, because other people may be trying to suppress you or shape you. Those who break through stand up for themselves. That becomes something that's admirable once you become successful, but the journey there is crazy. Just imagine—if we didn't have our individual characteristics, the world would be flat and boring. Which comes back to knowing someone else's settings and sounds. It's no good without the individual talent.

DAVE: It's the same with Herb's profession. You can read every book ever written about how David Geffen or Irving Azoff have done things, and then you can hang out your shingle as a manager. But that doesn't mean you're going to do what they do. There are so many factors that come into play besides what you see others do.

HERB: That's the beauty of creativity, across the board. There are always individual characteristics. And if you're fortunate enough to break through, and people respond to those individual characteristics, then really, you're kind of being paid to be yourself.

DAVE: Let me provide a metaphor. Let's say I run down to the Louvre and I look at the Mona Lisa, which, if you put a price on it, is probably hundreds of millions if not billions. So I set my easel down and I make the most perfect copy of the Mona Lisa that you could ever see, and I go to sell it, and it's worth nothing. Well, that's true in every profession. If people copy what I do and then they go to sell it—it's still just a copy.

HERB: And that is really the point of the magician's code. You can break it, but you can't really use it. Because even if you can emulate it perfectly, you can't create the real thing.

THE RANDOM ELEMENT

HERB: Sometimes there truly is a completely random element in achieving success. I've sat in label meetings as a manager, realizing that, in many ways you have no control over what will be successful, what will be the single, whether the promotion money is going to be spent. Those are all X factors that have nothing to do with your work.

DAVE: Let me modify that, because I slightly disagree. If you open a restaurant and your concept is to sell food, and you have pretty good, but not great, food, like say the Olive Garden, and you put enough money behind the operation for things like marketing, you're probably going to do okay. That same quality of food for a little mom-and-pop might not do okay, because they can't be just average and attract customers. But the thing that will give you the best chance of success is to have the best food at the best price possible. Still, it's never a slam-dunk. The label has to present it to the public and the public makes the decision.

HERB: The reality is that decisions are made that engineers aren't involved with. Someone else makes the decision about whether to spend two million dollars on this recording.

And you can spend that two million on a single and stiff. It probably has nothing to do with the mix; it could be some random factor that causes it to not happen. So, really, everybody has to work at their best and pray. Because there can always still be an X factor that can't be defined. If everybody could put their fingers on it, everybody would be successful. But it doesn't happen that way.

And it happens even less now. It was actually more structured when there were distribution points and radio points that were more closely controlled. All of that is much less so now. A lot of things had to come together, and a lot of people had a stake in the project. In a way, that made the magic of what happened in the studio even more important. Everybody involved was looking for that extra special sauce that would combine with the other elements to make a hit. That's why if somebody gave away secrets, or what people saw as secrets, some people were aghast. "How could you do that? We're already working in an arena where we have to hope things happen. If you give away some of the magic, that's going to hurt."

Now, there wasn't a big furor about it—that would be too strong a word. There were some people who reacted. But in some ways the furor was mitigated because Dave was so well liked. It wasn't some unknown villain who was sharing secrets; it was a beloved member of the fraternity.

DAVE: Back then nobody phoned anybody because we were all working together all the time. There was a high mobility; studios had four or five control rooms and over the course of a month you'd see everybody in the business. New York and L.A. were tied together with only a five-hour flight, so within the first few years I was out here I had already become friends with many of my contemporaries. And out of them, I'd say 10 percent just told me flat out to stop doing it, and another 40 percent would make a joke about it and I was supposed to read between the lines, "Hey, man, what did you give away today?"

That wasn't just somebody saying, "Stop!" But it was someone showing they were uncomfortable with it. And that was their prerogative. I have no problem with that. I had no problem with them telling me back then that I should stop doing it. I did consider backing off a little bit, but it didn't make any sense to me.

I think what sharing does is help give people an ear and an understanding. Knowing what goes into the whole thing actually raises the value of the work. It doesn't devalue the process; it enhances the value of the process. The more they know about it, the more they are going to appreciate it.

Herb has said that no one person can predict a hit. Nor can one person always create a hit, because a hit is something that appeals to people on an emotional level and everybody's emotions are different. You don't know which emotions they are going to like. So if you present them the exact copy of a previous hit, they've already heard it. And they're very likely not going to like it. They need something fresh to get their attention.

Trying to deconstruct what part of the process people are reacting to and purchasing, I'd say the more people hear something, the more they like it. Familiarization is part of the process of getting people to like something. But it can't sound just like the last hit. Part of the special sauce is adding a little something different that catches people's ears. That said, I don't think anybody ever bought a record because it was EQ'd correctly or mastered correctly or was recorded at a certain studio and not in a bedroom. People buy emotion. They buy a feeling and a vibe—three and a half minutes of escape from reality. So I think that when people say sharing

EQ settings is going to affect their career I would argue that they have never put their heart and emotion into trying to figure out the parts of a song and the process that people are really relating to.

It's idiocy to think that a person buys a record because it's EQ'd in a certain way. Now, if the EQ enhances the emotion and the feeling, then you've done something.

HERB: The correlation to this book and our overall process is this: We are not necessarily teaching something new. What's new is how we teach. People get the information in an emotional way and make a connection to the teachers and a connection to the format that relates to their own way of perceiving things. That helps them make the knowledge their own. That's why we put so much time into how we present things. You can find the information anywhere. It's the presentation that makes the difference.

What we are doing is making personal understanding widespread. It's that personal connection with all those people around the country and the world that that our fans can't have any other way. Those communal learning environments don't exist in the same way anymore, and people are hungry for them. Previously, there were only ever a very few that could get that personal connection, and now we are making it available to millions.

Something I'm proudest of with the show, which is an extension of the salon, and also an extension of our careers, is that we have so many people who started with no connection to recording and mixing who say, "I like this. I find it interesting." That's the secret sauce. It's not the information. You can get the information anywhere. I can go to to the Web right now on my phone and find the same information. What keeps people coming back is that extra connection. Without it, you're not going to have anything.

DAVE: The ITLs are an example of this. The concept is to share information by showing it, explaining it, and encouraging people to interpret it themselves and use it in their own unique ways.

HERB: The idea of breaking the code relates to the show in that it's an amalgamation of what we've learned and how the tools for communication have evolved. Today the tools allow you to be honest in a really scary way where everybody can shoot at you when you put yourself out there. Before, it was who had Dave's phone number. Now, we can hear from the globe two seconds later.

DAVE: I can tie this up with one concept. It turns out that we all know the code because we all know the magician didn't really saw the woman in half. So there really is no code. Giving people information about sawing the woman in half isn't the code; the code is the illusion of making people think you sawed her in half. That's what scares engineers: it's not really about giving something away; it's about destroying the illusion. Because we all know the woman didn't get sawed in half, but there's an illusion that the magician knows something that we don't.

INTO THE LAIR, NUMBER 72: CREATING A HUMAN FEEL WITH DIGITAL TRACKS

DAVE: I want to talk a little bit about feel. We're going to use a hi-hat track as an example, but you can do this with everything. In various ITLs we've tried some of these techniques, but now we're going to combine a few of them to manipulate the feel on a track.

If you ask drummers to play without using the hi-hat or a ride cymbal, they'll probably say, "That makes it a little bit more difficult!" Of course, the greats can play anything—they can play with their feet, left-handed, right-handed, whatever they've got to do. But when a drummer plays, he tends to keep the rest of his body in time by hitting the hi-hat. That's the reason that the hi-hat, shakers, or anything in that frequency range is so important on a track. Sometimes, when I'm working with programmed drums, I'll have a drummer come in and add a live hi-hat part. That can make the whole track feel like it was done live.

There was a drum machine back in the day called the Linn 9000 that everyone used because it had the greatest feel in the world. I talked to the guys that worked on it—Roger Linn was one of them, of course, but the Forat brothers, who were famous in L.A., did a lot of work on them also. When I asked them why the 9000 felt so good, they told me that it was because it had a very loose clock and the timing was actually moving around. Not in any special order, just moving around. And that's why we all fell in love with that drum machine.

That got me thinking about how I could make a track feel a little different by manipulating the timing and giving every hit a different pitch. I used a track by my friends in the band Follow Your Instinct for this ITL. I just used the drum track, which has a hi-hat part that is very tight—dead on the money.

I made a copy of it and muted the original. In Pro Tools you can use the Random Pencil Tool to do this. I selected a sixteenth note and checked it out with some random volume to see if it helped the feel. Turns out it's pretty subtle, but I noticed a little something.

Then I messed with the frequency a little bit—automating a very tiny amount of frequency change. I happened to use a plug-in that has an up-and-down setting built in, so I put that on a track. Now we've got our volume change and our frequency change, and it feels a little better.

I also shifted the entire copied hi-hat track two milliseconds early so I could randomize a delay—anywhere from zero to four milliseconds. That means sometimes my hi-hat will be ahead two milliseconds, and sometimes behind two milliseconds. I also automated the delay. Now, that's kind of cool! You can feel it. We've got some little artifacts, but this is just for demonstration purposes and you can tighten it up later.

So we've randomized and humanized the volume, we've kind of humanized the pitch of the sound, and we've made the timing move around a little bit. All of those things combine to make this track feel more human. Not perfect, but human.

Another thing you can do is run the two together—the original, perfect one, and the imperfect copy. In this case, that turned out to be my favorite combination.

We can use these same techniques with vocals, or with any instrument we want, and we can take this to even more levels depending on how long we want to manipulate. If you have to draw this in manually, it can take you a while. But you can copy and paste once you've completed a bar. I'd recommend two bars, just to get a little more randomness to the feel. Pro Tools, and I think all the DAWs, have a randomizing pencil tool, so check it out. Experiment with ways to make things random, knowing that very small amounts of randomness are the same as staying human. Because none of us are perfect. And I'm at the top of that imperfect list!

5 Home Plate

How the Legendary Studios Shaped Me

DAVE: When I first moved to L.A. I wanted so much to be a part of the record business that it sometimes hurt. I read *Billboard* a lot. I read a recording magazine that's been defunct for a long time called *REP* (*Recording Engineer & Producer*), and I read *Mix* magazine. I was a credits fanatic, so I knew every studio, every producer, every engineer, every record company, and most of the A&R people by name, and a lot of them by face because of the pictures in the magazines. So when I first got to L.A. and I wasn't working, I started going around to different studios and sneaking in, just to see the studios—and sometimes to grab a little bite to eat!

Larrabee somehow symbolized the pinnacle of the profession to me—I felt like it was my portal into the business. Now, you have to realize, at that point in time success was not a goal. It was too far away. Just getting a foot in the door was my goal, and I believed Larrabee was the place for me to do that. So how I got there was . . .

After I started working with Wolf & Epic, we were editing Bell Biv DeVoe tracks at Wolf's home studio. We'd mix the tracks at pretty nice studios, but they weren't Larrabee, so I talked Wolf & Epic into letting me edit one of the songs—I think "Thought It Was Me"—in a closet at Larrabee. When I say closet, I mean that literally. As you entered through the front door of Larrabee West (which at the time wasn't West—it was the only Larrabee!), the receptionist would be at the front desk, and immediately to the left was a little storage closet. They rolled two 2-track machines and a couple of speakers in there for us, and that's where I did all the edits. Now I was working at Larrabee!

That same year, [Larrabee owner] Kevin Mills invited me to their annual Christmas party and gave me a Larrabee leather jacket. I honestly teared up when I got the jacket. Getting that jacket made me feel like everything I had gone through up to that point had been worth it. I was making a little money, no doubt, but symbolically, it was working at Larrabee and becoming part of that family that made me feel like I had a USDA certification stamped on my forehead.

LEARNING TO MIX

Before I got to Larrabee, there was a period of time when I was working at Aire L.A., which was owned by Craig Burbidge. That was also an extremely important part of my career, because Craig, Chuckie Booker, and a bunch of the other guys there really made me feel like I was a professional. I know I keep emphasizing the idea that I needed certification that I was okay, that I was part of the club, but I think it is in some way important to everyone. To work hard without feeling like you're part of a particular group is a lot more difficult than to work hard and feel like you're accepted by your peers. That acceptance helps you compete a little better and in a healthier way. I know that because when I first moved to L.A. I was out to clean the clock of all my heroes. But when I actually met my heroes, I just wanted to be as good as they were.

Now when I say clean their clocks, I don't mean anything too negative. I just wanted to be better! Meeting my heroes made me even more competitive. But the competition was more like a friendly game of cards or going fishing and vying to catch the biggest fish—as opposed to catching all the fish! Meeting my heroes also made me more humble. It put a face on the competition, which made it not only healthier but a lot more fun. It also made me work even harder. For 15 years it was very difficult to pry me out of a control room.

At Aire L.A. I learned a lot about processing. When I first got there, I didn't know you could put processing on the stereo bus. My assistant, Anthony Jeffries, asked me what I wanted on my stereo bus and I said, "You can do that?" He started laughing, and I tried to cover up with "Oh, I'm just teasing, man—just messing with you." Then, of course, I asked for whatever Alan Meyerson used, and that's what Anthony gave me. Whatever Alan used, that's what I wanted to use.

There was a little stretch at Conway that was pretty cool also! I got to work with Ringo Starr, and I also met Mick Guzauski there.

So it all was starting to get glued together, and it solidified for me when I got to Larrabee. Arriving there somehow made me feel like I had moved from racing go-carts to racing NASCAR. Going around a track is going around a track, but it felt like I was doing it at a higher level when I was at Larrabee. Not because I was at a better studio; my control room at Aire L.A. was one of the best I ever had. It was something else. Pick your own metaphor: if it was baseball I was in the World Series; if it was NASCAR, I was at Daytona.

Remember, early on it seemed like every time I looked at the credits on a record I liked, it had been done at Larrabee. So it had to be a special place, and if I could just make it to Larrabee it was one less variable I had to worry about in the process of getting to where I wanted to be. That was my whole reference for the records I liked: who made them, where did they make them, what was the label, who was the A&R person, who was the producer, who was the engineer, and who were the musicians? If Babyface was the producer I would automatically buy the record. If I saw Jon Gass, Dave Way, or Steve Hodge with Jimmy Jam and Terry Lewis, I bought the record. Because I just wanted to see what they were doing. Eventually they became my contemporaries, but prior to that point in time I was like a young kid playing basketball on the playground, pretending I was playing in the finals at the NBA. And then, because I was at Larrabee, I was on the same court that the finals at the NBA were being played on. I still wasn't in the NBA, but it felt pretty good.

ON TO THE ENTERPRISE

I really learned fully how to mix at Larrabee, and I had a good long run there, but eventually I wanted a change and a new creative environment. Then an opportunity came along for me to move to Enterprise, where a lot of engineers were in residence. The Enterprise studios had been built by Craig Huxley, a child actor who had appeared in the *Star Trek* TV series, and who later became an Emmy-winning and Grammy-nominated keyboardist, inventor, and film producer who also composed music, including for several of the *Star Trek* movies.

At the time I moved there, Enterprise was the heavy metal, hard rock studio of choice. All of the corporate-type hair bands worked there. They didn't really have an R&B/hip-hop kind of room, so I brought my clientele there. There were five control rooms in the building, and we pretty much turned the place into party central.

It definitely got crazy and there was always something going on. The main ringleader was producer Damon Elliot. One day he and songwriter/producer Linda Perry filled the studio parking lot with snow. It was the middle of summer in Burbank and they rented a snow machine from a ski slope and filled the parking lot with several feet of blown-in snow. There were snowball fights in the hallways of Enterprise! I'm trying to mix and getting hit in the back of the head with snowballs!

All of my friends were working there and we had a lot of fun. The real outlaws there were 2Pac's group. They were there all the time and they made it interesting every day. Alan Meyerson was working there; I met Andy Wallace and Jay-Z there, and so many, many other people. It was the social hub for the engineering crowd. My control room was the first control room you saw when you came into the building. I always left my door open, and everybody who came in would stop by to say hello. It was a very free-flowing time.

It's funny that in Atlanta, no matter how good I was, I could always say that if I had been in L.A. and had everything all the hot dog engineers had to work with I could have been better. When I found myself in the hot dog capital of the world with all the equipment, I had no excuses anymore. I had to be better, because the only difference between me and the guy in the control room next to me was me! I could no longer make excuses that I didn't have the best equipment or the best facility . . . now it was on my shoulders to be great. Oddly enough, though, those guys in the next room really helped me. They were friendly, and they pushed me. Gently, but they pushed me.

That scenario also translates to today. Engineers who have been doing this a while get more respect than ever. Because now that truly everybody has pretty much the same gear, everybody understands how good the good people are. And I say that humbly; I am not saying that arrogantly. Someone can have the same plug-ins, the same DAW, the same speakers, the same everything! But they don't sound as good as the great engineers. It's like, everybody watches Tiger Woods on TV. If he says he uses this piece of gear, people go out and buy it. But they're still not as good as Tiger Woods, and they know it. And that makes them realize how truly great he is.But the gap between Tiger and the guys trying to be Tiger is much bigger than the gap between me and the guys trying to be me. There are a lot of guys out there working at home right now who are really great.

A sad thing is that in my musical career it seems like I ran into a lot of people who said, "If I don't make it by 25, I'm quitting." A lot of the true heroes in music left early to take care of their families. They left early because they wanted to be able to pay their rent and because they didn't want to go through a lot of the struggles that Herb and I went through. We often talk about those days as being incredibly fun and wonderful, but there is no way we would go back to them ever again. In important ways they were not at all fun, and even knocked a lot of greatly talented people out of music. Of course, that happens in any profession—art, pottery, writing. . . .

THE MECCA THEORY

HERB: One of the great things about Enterprise was what I call the Mecca Theory. Recording studios in that particular time were (and still are) Meccas, really magnets, for everyone: engineers, artists, producers, label people. Studios drew us like moths to a flame.

A lot of what differentiated those studios and made them Meccas were the central personalities you found there. You knew you would see certain people at Enterprise. You recognized their cars in the parking lot, and you knew if you were there you'd see A&R people coming through. You conducted business, but it also became social; it became family. And what was great was that there were a fair number of Meccas.

Now there are too few Meccas, but for those that exist, like the Village, Record Plant, and others, it is still about the personalities you are going to meet there. I know with the right equipment and engineer I can get a great mix anywhere, but I also want to get that other thing, where you can meet and get work done. That creates commerce and relationships and places where people learn.

DAVE: Which is something that was very germane to a person's professional development. The new version of that, of course, is the Internet.

HERB: This is something we are trying to address with the show, because the new version, the Internet, can miss a lot. In some ways it creates a false connectivity. Instead of real interpersonal relationships, you just have communications. What we try to do is make the Internet more personal.

The show is a meeting place for creativity and ideas and talent. You could say that we inject an analog form of interactivity into a digital layer of platforms. That's the secret to the show. It takes advantage of the current lack of interconnectivity.

It's a phenomenon today that people could not be more connected, yet they are also more disconnected—all at the same time. The Internet provides information, but it doesn't give you perspective. You don't get to learn from the masters; you don't get to share and put in the hours together.

The trade-off, of course, is that you get tools that were never available before. So of course it's not all bad. But most music sessions today don't approach the experience you have when you are physically present mixing with people in a studio. You can send them the files, you can talk, you can Skype, but it's not the same as stopping by the studio, spending time, and having a conversation. You can get the work done, and we are the beneficiaries of that—we take advantage of that fully. But there is an experience that you get in anything you do, and mixing remotely is just not the same experience.

I can date on the Internet, and I can even have Internet sex, but it's not the same as getting to know someone in person. John Travolta can buy a plane on Craigslist. You have access to just about everything. But at some point you are going to have to touch that plane and actually fly it.

Disconnecting from those experiences is, I think, what a lot of today's audience gets wrong. They think, "If I have the gear, and I learn on the Net, I'll get there." And that can take you to a point. But then you have to search out how to get the real interconnectivity. We see that a lot in our world when we get inundated with requests from people who want to pay just to come and have time with Dave or with me. They don't even want a mix. They just want to sit there and to shadow us.

DAVE: I have worked at home, but I have found that I need a big control room with people in it as much as my clients need a big control room with me in it. That's the symbiosis of the process. If you really, really want to make great records with your clients, you kind of need your clients around you as much as possible.

Skype helps, sending pictures of the artist helps, talking to the artist helps. Herb mentioned earlier the importance of the process of talking to clients in advance and understanding their needs so we can figure out if we are the right fit for each other. What works is to have as much human contact as possible.

I differ from Herb on one small point. I think the old studio system was great for making connections, and I think the Internet is equally good for making connections. I think the difference is in the skill with which people maintain their relationships after they've made that connection. People-to-people skills were quite a bit better in those days, and it seemed like people turned those connections into something more than it seems like people are able to do today. I'm the beneficiary of many of those connections getting turned into something real. Today, a lot of people don't communicate in ways that lay the groundwork for a continuing conversation.

THE GRIOT THEORY

HERB: Back in the day you didn't have a choice. If you didn't create any personal skills to help get to the next level, you were stuck. People actually worked on developing those skills. I liken it to what, in my world, was part of the evolution and development of executives, and I call it the Griot Theory, after West African storytellers called Griots. It's a tradition there that kids have to sit and listen to the Griots, who are the keepers of the culture and history of the tribe. I also had to do this coming up in the business, and, growing up, black families in black neighborhoods did this a lot.

Griot Theory grew out of recording sessions. As a young executive, I'd be at a session with a drummer like Steve Gadd and all these incredible musicians, and my bosses would say to me, "Just sit!" So I did. I'd sit there while they worked and just listen to those guys talk until the sun came up.

The studio system back then had a great leveling effect—once you got in the gates. Clive Davis could be there, and if you were in the room, all of a sudden you'd find yourself sitting with him. So as a young guy, I'd find myself there thinking, "Oh, my gosh, I'm sitting in this session, oh, my gosh, just don't move, don't make yourself noticed, don't have anybody throwing you out!"

"Genius needs to be sautéed."

Once you got to be a veteran and respected you would see young kids sitting with you, hoping you wouldn't throw them out. There was a development process that happened naturally where those kids learned that you had to have interpersonal skills. You had to learn from others or nothing would happen for you.

Today, you can move up the chain and never come out of your room. You pick up some skills and even consider yourself mixing, but you can't get into the pantheon without knowing how to deal with people and all of the psychological elements, and the management of people and personalities. It is harder now to acquire those skills, and we get requests from people all the time who want to know how to do that. They recognize in their gut the need, but they don't know where to go or how to learn these things. So we are trying to help them learn some of it from the show.

We also get the questions a lot, "Should I move to New York or L.A.? Is there a studio where I can intern?" And these days we don't have that many options to refer them to, so it's hard to replace that necessary skill set.

HAVING A SOUND

DAVE: It seems to me that the engineering process is complicated by the fact that most guys, when they get to a top level, are capable of doing just about anything—except reading minds. If I could read minds I would be completely efficient at giving my clients what they want. Now, back in the day, it often happened that you hired an engineer, not because you really wanted him to read your mind, but because you wanted him to do what he did. The skill set that he brought to the process was what you needed at that point to take the song to the next level.

Over time, that concept became somewhat obsolete, because you didn't always hire engineers for their sound anymore. For many of those engineers their sound had become irrelevant as styles changed and moved on. So you started hiring engineers for a new sound, or to get your sound. That split the profession somewhat. There was a new group of engineers who were hired for their specific sound, and there was a group of engineers you hired so that they could help you get your sound. Now, I maintain it doesn't have to be one way or the other. As an engineer, why not do both and be flexible, depending on the needs of the client?

I've noticed that the process for engineers now is that they are finishing the mix part of a production as opposed to starting the mix where the production ended. This is because productions mostly don't end these days; they just morph into the mix stage. That requires a different mindset from the engineering crowd. You don't want to inflict your hard and fast concepts onto the mix at that point, because it will destroy the vision. And you also can't have two visions. Any creative process that has two visions is probably diluted and doomed from the start. All creative processes need one clear, almost dictatorial vision. That's why, in general, artistic endeavors that are tenaciously monitored and held onto by the artist, as opposed to the publisher, the A&R person, etc., tend to be the ones we remember the most.

In my formative years, the music business was run by an industry that had great power to determine what reached the marketplace. Now, thanks to the Internet, we're in a period where there's no real control over what reaches that marketplace. And therein lies one of the big problems of the Internet. In a way it's a good problem, because everybody has a chance to be heard.

HERB: Let's be real. A lot of us were drawn to show business by the magic. It was powerful, sexy, you could get rich, and women loved guys in show business. Sometimes that magic extended past the artists to executives. Even better! Lawyers, managers, and executives became focal points, too. Today, who is bigger, Simon Cowell or Psy? Clive Davis or Jason Mraz? Russell Simmons or Ed Sheeran?

The other magic is getting to work with genius. If you are lucky enough to find yourself in that position, it is life affirming and life altering. Spend a couple of days in the studio with Quincy Jones. Sit and watch Pharrell produce all night. Be in the control room when Whitney Houston is singing or Diane Warren is writing. Watch a kid named Tricky Stewart in his lab while he is creating, or Mary J. Blige producing vocals on your artist. It elevates your sense of what can be, how you think about creativity, being fearless. It is instructive, rare, and defining.

It doesn't matter if it is to your taste or not. It could be Lumineers, it could be Bieber, it could be L.A. Reid's hand in some project, Dave's mixing ability, or Scooter Braun's business sense. I love Troy Carter's work! All of these share the ability to touch talent and inspire, to see the bigger picture and appeal to people.

The Internet has sort of flattened out that process. You do stuff from the house and you do it on Skype and it doesn't quite create the magic. Quincy Jones once said, "You will never come into my house and see a recording studio." When we asked why, he said, "Because the studio is a church. And when you go to the studio and you are going to church, sometimes God comes in. God doesn't just randomly stop by your house." So you always want to be working in the kind of place where God will come in. Quincy's recognition is that the magic and the spirituality and the specialness of the studio are what make it a creative environment.

BRINGING THE ASSISTANT ALONG

DAVE: You spend a lot of time with your assistant, and what makes that relationship wonderful and unique is the total trust you have to share on every level. You see your assistant more than your spouse, more than any other person in your life. You need to pick that person carefully. There has to be personality compatibility, and of course each individual has a very unique personality. They have to be 100 percent self-starters who are able to constantly think on their feet. But they still have to be subservient to both your business and your philosophical needs. And in my case, they have to be amenable to being manipulated behind the scenes by Herb!

HERB: Because Dave is a philosopher, it's been fascinating for me to watch him insist that all of his assistants bring their own philosophy to their work. In order to work with Dave, they need to develop a point of view that attaches to their craft. For example, Jaycen's was edgy, Dylan's was thoughtful, Ariel [Chobaz] was more musical, and Drew [Adams] took a street approach. None of them shrink from expressing their personalities and perspective, and all of them have emerged very strongly in their own careers. Of course, they also developed the skills to know when to interject their perspective and when not to.

Dave didn't always agree with them, and I almost never agreed with them! But they recognized the importance of having their own position. I think that's different from how a lot of other assistants get trained. The others may get trained technically, they may learn how to deal with clients, but it's not necessarily desired that they have a point of view. Dave's assistants have to bring something to the party. But they also have to be subservient on some occasions. That's pretty high-level stuff to balance

DAVE: It's difficult math, because there are multiple jobs. It's like being a sous-chef, the person who chops up the vegetables so that the chef can put the dish together. Also, the job requires being a creative companion. So I train my assistants to be a big help on the creative level as much as I teach them how to print mixes and do all the mundane stuff.

HERB: That's where the philosophy comes in. They are engaged in the process early on, even in the way you introduce them to the clients, to artists and managers—they are always introduced as a valuable member of the team. And they have input. As a manager coming with projects that had platinum pressure, I felt comfortable knowing that Dave's assistants were part of the process. They weren't just a person who could get me a Diet Coke.

It always felt more like a big team, and that I, and my artists, were in good hands. If I couldn't get to Dave, but I could get to his assistant, I knew the assistant would give me the proper information and handle correctly whatever I needed done. For managers and producers, that provides a basic sense of confidence about the team. I'm not surprised that those former assistants all have emerged very quickly in their careers. Once they left Dave, they were ready to take off.

DAVE: I've always looked at the record-making process as a team sport, and my team starts with my manager, in this case Herb. He's the first exposure to the team for most people. But everybody has equal importance, starting with the receptionist who greets the clients when they come in—because if that greeting is really nice, my mix just sounds better by the time they hear it. If the greeting is not nice, my mix is going to sound different. If you're a record company president and you are treated with respect at the door, and when you get to my room the respect continues, it's just a much easier process.

I liken the assistant education process to what happens to a person who wants to be an artist, a painter, and who chooses to go to school to learn the craft. Let's say this person has a very unique painting style. Odds are the formal institutions and places of education where this painter would get trained would probably steer him or her to the taste of the teachers. And when he came out of that institution, he'd no longer be that unique person and would probably have assimilated too many of his teachers' influences.

In contrast, I've always felt that assistant engineers have more value to me if they become the best at what their hearts want them to be, as opposed to being a clone of me. If you survey the landscape of all my successful assistants, not a one of them sounds like me. They sound like themselves. And in the process of developing their own style while they worked with me, they also became more valuable to me.

I don't need a clone. When I turn to the assistant and say, "What do you think about that snare sound?" I don't need someone who says, "Dude, you're a genius!" because he has the same taste that I do. I want conflicting, unique taste. I need to be able to ask, "Is this vocal too bright?" and have him say, "A little bit." I'm blessed that I can do that. I don't need to do it too often early in the day, but when you're working long hours, having an objective opinion can be helpful. And I can delineate a few times when I believe that concept has actually helped sell records.

It goes back to my theory that, if you're trying to make a record that millions of people want to hear, and you start by pleasing one person—yourself—mathematically that has a certain probability of succeeding. But if you please two people, you've cut the odds in half that you're going to be successful. You've doubled your chances. If you please four, you've quadrupled them. And if you get a roomful and you use them correctly, it should greatly increase your odds of succeeding.

Of course, at some point it becomes diminishing returns, because you can't please everybody. But it is probably not a bad idea to train an assistant to give you good advice. It's also helpful that, at some point, usually about month three, they start caring about my personal health. I've seen that my whole career. They'll say, "Dave, um, we haven't been sleeping a lot lately, are you okay?" Or sometimes you'll forget to eat, or maybe you've eaten cheeseburgers five days in a row, and they'll say, "About time for a salad, don't you think?"

HERB: That's part of how the Meccas take care of people, because it is germane to the operation of the whole shop that everybody is taken care of—whether it's the people in the studio or the fact that when you pull up to the gate, if they know you and you're part of the family, the gate just opens. The Meccas are really incredible petri dishes. So many things get birthed there.

DAVE: As the university system is to academia, studios are to music. The first thing I did when I moved into my new facility was send out pictures, which was my subtle way of saying, "Drop by!" Most of my friends dropped by within the first two weeks.

HERB: There's nothing else in the music business quite like that. Where else is there to go that's communal, where you go to see other people and you also get inspired? You can't do it at rehearsals, because they're fleeting. They come and go. You can't do it at a record company's corporate offices. It's that spiritual component again—that's probably lacking in those offices. And there's an exclusivity component, too. It's a country club where you have to have membership.

DAVE: You have to be certified as part of the club, and the way you get certified is just by doing the things that most of the club members do.

HERB: You got granted access. You couldn't just come and show up. You had to be granted, and once you were, you could have the grant pulled. So ultimately those who were pros made sure they kept access and contributed. And it was completely fulfilling when you did. I don't care what part of the business you were in; if you went in and spent time, you were elevated when you left. You heard good music or met people or made a connection, or just learned something!

There are still some Mecca studios left. But every process evolves, and in some ways, the Internet serves that purpose now. If you take the Larrabee/Enterprise concept and what it meant, and move forward, there is a correlation to *Pensado's Place*. The show is that community. The people who reach out to us, their desire to get into the studio and get near that, still holds. It is still powerful even in the Internet age. "It's cool, I have all the tools, I can read about it, but can I get in the studio, how do I get in the studio?"

DAVE: The Internet has changed the restaurant business and it has changed the music business, but there are still basic concepts of both that will never change. You need to have quality food. You need to have talented people to prepare it. You've got your choice of fast food, your choice of $200 meals; there's a variety of choices. You can do your research on the Internet, but until you walk into the restaurant and sample the food you're not going to understand how good that chef is and what the process is really about.

And you still have to interact with the people who can get you to the next level. A great engineer, not just a mix engineer, but a great engineer, still has value to most people. I've had hundreds of successful records and I still find comfort in playing my work for other engineers in order to get their opinion of my work. It's interesting, though—mix engineers can be kind of funny. We don't give more information than we are asked to give when we listen to somebody's work. So if I go in and ask, "Is this any good?" I'll get one opinion. And if I go in and say, "I nailed this! Best thing I've ever done. What do you think?" I'll get a different opinion.

From time to time I still need, and I think we all still need, positive feedback. There's just no greater feeling than to think you've done something really good. When I played in bands we'd get instant gratification and evidence that we were good. First, you come out alive—that's a good sign, when you leave the club. Second, you get hired back. And immediately, people applaud and cheer and stand up and you feel the dance floor.

I miss that severely in the world I'm in now, because sometimes I have to wait months before I know if what I've done translates to millions of people. But there is that little tiny shortcut that helps, and that's when you play it for your peers and for people whose opinions you respect. If you play your stuff for people who are always going to tell you it's great, you may as well just forget it!

The legendary studios accelerated the process for me. Because of the way I'm constructed, I've got the worst combination of arrogance and insecurity that probably anybody's ever had. So it forces me to do things a little differently than other people. Arrogance by itself probably produces better results. Herb has a concept that he's taught me over the years about how some groups of people get successful. He uses it in business meetings, but it can apply anywhere. Well, he doesn't call it arrogance; instead he's always telling me I should do it like the rappers do it. Explain, Herb, your concept of how people can push the right way and get what they want without being perceived as irritating.

HERB: Dave is a philosopher who is also very respectful of everybody's feelings. That can get in the way of him being direct about getting business done, and sometimes people will take advantage of that. He's aware of this, and I think sometimes he enjoys watching me be direct as a kind of sport.

It's my belief, and it's also the way I was trained, that cutting specifically to the chase does not mean that you have to be brusque and irritate people. The rapper correlation is that people in the hip-hop culture can be very specific—as in direct and cutting to the chase to ask for what they want. But in many ways that means that you are being mindful of people's time. In business, your dealings with people are generally goal specific. Everything is not a teaching moment and often the first person who has to explain something loses. You need to talk to people like-mindedly, assume they know what's going on, and cut to the chase. Time has to be respected, efficiencies are important, and everyone is trying to achieve something. Preamble, predicate, theory, and thesis all have their time and place, but generally that time and place is not a business meeting. Oftentimes, if you answer why, when you are being asked how, people do not pay attention to what you are saying.

Dave's easygoing way has tremendous value, but sometimes it can hurt him because people may take advantage. I only get involved when it harms his business. From pricing, to not paying on time, having unreasonable expectations, the "never-ending mix," non-payment, etc., the potential issues are many. Because Dave is so easygoing, they may do it in friendly ways, but it still hurts. Dave would "Aw, shucks" it away, and that wasn't good for his mix business. We needed to change that, and we did. But it's an object lesson for anybody up and coming. Be civil always, do good deeds, even give back, but always take care of your business.

RIDING THE INTERNET TIGER: THE THREAD TO *PENSADO'S PLACE*

HERB: In today's mixing world, the elite are dealing with higher expectations, increased competition, and compressed budgets. The need for efficiency is very high. Engineers are mixing major records, indie records, and records that come in from the Internet. They are operating in all kinds of environments. Since that's the case, sometimes you can find yourself busy as hell but not earning the living you need to earn. Therefore it's important to be focused and active with a game plan. Working smart is critical. Following the game plan and adjusting accordingly is just as critical.

A fun part of our journey with the show is that it has evolved our business relationship as partners. It's now more than just the singular focus of managing an engineer and booking engineer gigs. It's a different, more complicated business, with so many things that intersect. We're maintaining a weekly audience and have to do all the other accompanying things. It's a tricky balance. We had no idea we'd be so successful, so now it's forcing us out of our normal habits, which makes it interesting and stimulating and frightening every day. It's actually very strange. I've had this weird career, I've managed all sorts of things, and now I'm an also an act! I'm managing me! Everything now has a different kind of impetus to it, and the pressures have changed. What we strive to do is evolve, without losing who we are.

DAVE: Not only are we now "the artist," but we're unique in that we're the artist because of what we do, what we did, and what we will do. In other words, had Herb not been a successful manager he wouldn't be a success on a TV show.

HERB: It's wonderful that the place that Dave and I inhabit today allows us to speak to up-and-coming folks about how you do it today—not about how it was done yesterday. We can talk about the fundamentals of mixing from a true master's point of view, but also about how to be a content creator, publisher, practitioner of new media, and global communicator. Plus we have to monetize! It's a lot to combine and correlate. But we are living it, not theorizing it. And that's a very different, very exciting thing. We've learning every day with our "side" gig!

DAVE: Audio is the foundation of *Pensado's Place*. But the show is really about creativity.

INTO THE LAIR, NUMBER 54: THE TOP 10 MIX MISTAKES

DAVE: I've been asked to judge several mixing competitions, and during that process I've noticed some areas that continue to need improvement. There are definitely places where we can all step up our game.

Now, it could be said that I and all of my buddies—Tony, Chris, Manny, Jason, Dylan, all of us—have to enter a mixing contest every day. Sometimes it's even an actual mixing competition! I've had situations where the record company didn't tell us that they were sending the mix to four different people. So you never stop entering mix competitions. We have to do it a lot. And while listening to some recent competition entries, I've come up with 10 things that seemed to stand out as generally needing work.

1. Energy and Emotion

   I'm a stickler for energy, emotion, feeling, and vibe in a song, and I notice that the mixes that I tend to choose all have more of those things than the others. One way to get them in your mix is to choose a speaker to monitor on that takes some of the hype factor out. Some speakers are tweaked to add highs or lows to sound more appealing. That's fun, but it's not the kind of speaker you want to work on. You need to mix on speakers that have a flatter and more accurate response, because you want your mix to sound good on every speaker. So maybe check your mix on a laptop speaker or some Auratones. Monitoring at certain points in the mix at a slightly lower volume can also help take out the hype factor. If you can make it sound loud and exciting at low volume, it will sound loud and exciting at any level.

2. Determine What Elements Are Important

   For a particular song I was working on recently, the most important elements were the vocal, the cool-sounding guitar, and the kick drum. As the song progressed, certain other minor characters came and went, but those three elements went through the entire song. They were where the money was. Those are the sounds that can't just be good; they've got to be unique and spectacular. They have to convey a lot of the energy and emotion. You need to figure out what those elements are.

3. Have a Focused Idea or Plan

I notice in the contests that people can't decide whether they want to be one genre or another: modern, old, analog, digital, classic rock—it was kind of all mixed up. You need to get a plan. As I've said before, it's not that you can't mix in some elements of different genres; you can, of course, and that can be very cool. But you need to have an overall plan for what you want the song to sound like.

4. Determine Where to Be Inventive

I've heard mixes that were super, super inventive, but the fact that they were inventive didn't contribute anything except to the mixer's ego. You've got to pick and choose your areas depending on the song. Make it meaningful!

5. Balance, Balance, Balance!

You don't want a spaghetti sauce that just tastes all garlicky! I hear a lot of songs that are EQ'd spectacularly, and would have no flaws except that there is something missing balance-wise. Just moving one instrument can change the balance of everything. You need to pay attention to that fact. When you lower the volume of the guitar, the drums may get too soft. When the guitar is too loud, the vocal may be too quiet. Don't set your balances and levels by anything you read in a book or were taught. Set them by what feels good and helps the emotion come across. Also, check the overall balance in different modes. The classic studio technique is the "outside the door test." Go out in the hallway and listen. At my project studio I'll go into another room and listen to the mix at a medium classic kind of level, like a home TV level. You'll find that helps the elements become easier to discern.

6. Panning

Now this kind of shocked me. We've discussed panning on the show! Several of our guests have talked about how they pan and why. We've gone over LCR. But there were a couple of songs I was judging that I might have put in my top five if they hadn't made my head go sideways. They just felt out of balance. Panning is a tool and technique that helps keep the listener's interest. In a painting you don't see everything at once. Your eye wanders around the canvas, and a good artist can control where your eye goes so you find different things in the exact order that the artist wants you to find them. In the audio world we want to do that same thing; we want to take control of where the ear goes and use that to our advantage.

7. Understanding EQ and Compression on the Stereo Bus

In a lot of the entries people had put something on the stereo bus that didn't really enhance anything. When talking about compression, a lot of mixers use the term "glue," and I think they're correct. You want compression to glue the mix together. You don't necessarily want it to make the mix sound louder; it just needs to make it sound better. Sometimes louder is better, but sometimes more dynamics is better.

I also noticed in some entries that people seemed to be EQ'ing just because they thought they were supposed to. Many of them were adding low end, which does absolutely nothing to help. So make sure if you EQ there's a purpose. Don't do it just because you saw your favorite engineer do it. Have a reason for everything you do, and make sure that reason is rooted in energy, emotion, feeling, and vibe.

Remember: The stereo bus is sacred territory. In general you want to use EQ and compression on it subtly. Don't overdo it!

8. Using Effects for No Reason

Another thing I noticed was effects and reverbs being used for no discernible reason, and chorusing and flanging that came out of nowhere. Again, whether they're subtle or radical, effects should be used to contribute to a feeling and enhance the performance. You don't want listeners to hear the effects. You want them to hear the performance. Make sure the effects match the concept you have for the song.

JAMES FAUNTLEROY: "I really do have eyes!" 420. Killer writer.

ERIC VALENTINE: Rock savant. Great guy.

JUSTIN NIEBANK: Nashville magician. Funny. Gave us early Country.

WES DOOLEY AND DAVE: The Mad Hatters.

DAVE AND JACK JOSEPH PUIG: Compadres. Leaders. Legends.

HERB, STEPHANIE "SPITFIRE" WILLIS, AND DAVE: Stephanie is our Nashville angel. Dig Dave's pocket square, which was from the table setting.

JOHN MCBRIDE: Southern force majeure. Excellence without compromise. Brother from another mother.

DANN AND DAVE HUFF: Country brothers extraordinaire.

HARMONY SAMUELS: One of London's best imports.

SALAAM REMI AND DAVE: Amazing musical prodigy. Clearly thinking "White boy, WTF?!"

DAVE, TONY MASERATI, AND HERB: Monster talent. Classy and elegant. Our brother.

JAYCEN JOSHUA: Gifted. Outspoken. Successful. Dave's protégé.

MIKE ELIZONDO: Diverse. Badass.

DAVE, HERB, YOUNG GURU: Producer-engineer with Jay-Z. Educator. Brilliant.

WILL THOMPSON: The Third Pensado Musketeer.
There's no us without him.

TYLER WARD AND DAVE: YouTube giant. Talented artist. Cool as hell.

DA INTERNZ: Kosine and Tuo—Da Internz. Fiesty, spirited, smart, funny, and gifted.

HERB, VANCE POWELL, F. REID SHIPPEN, DAVE, NEAL POGUE, AND CHARLIE PEACOCK: The Ticks or Treats panel, Belmont University, Halloween 2013.

NASHVILLE GIVES THANKS SPECIAL: Stephanie and John Willis, Herb, Ed Seay, Steve Bogard, Buddy Cannon, Dann Huff, John McBride, Martina McBride, Kevin Becka, Justin Niebank, Dan Frizsell, Dave, Tony Castle, and Brett "Scoop" Blandon.

ALEX DA KID AND MANNY MARROQUIN: Alex da Kid, star maker. Manny Marroquin, smash creator. Two at the very top of the game share a laugh at Gear Expo.

Pensado's Place Gear Expo at Vintage King in L.A.

LUCA PRETOLESI: EDM trailblazer. Visionary. One of our favorites.

CHONGOR GONCZ AND COLE NYSTROM: Pensado kids. Big futures.

MATT MAHER: Opened for the Pope. Spiritually uplifting. Crazily talented.

DJ ALI: Art and business. Self made. Love him.

BELMONT UNIVERSITY Q&A: Waiting to learn. Tornado outside.

COLE NYSTROM, DAVE, HERB, BRIAN PETERSEN:
Team Pensado chewing the fat.

SHEVY SHOVLIN: Our Vintage King brother. Shevy was an early believer. He's been with us every step of the way.

DAVE TOZER: New York's finest.

MARCELLA ARAICA: Miami heat. A studio wiz.

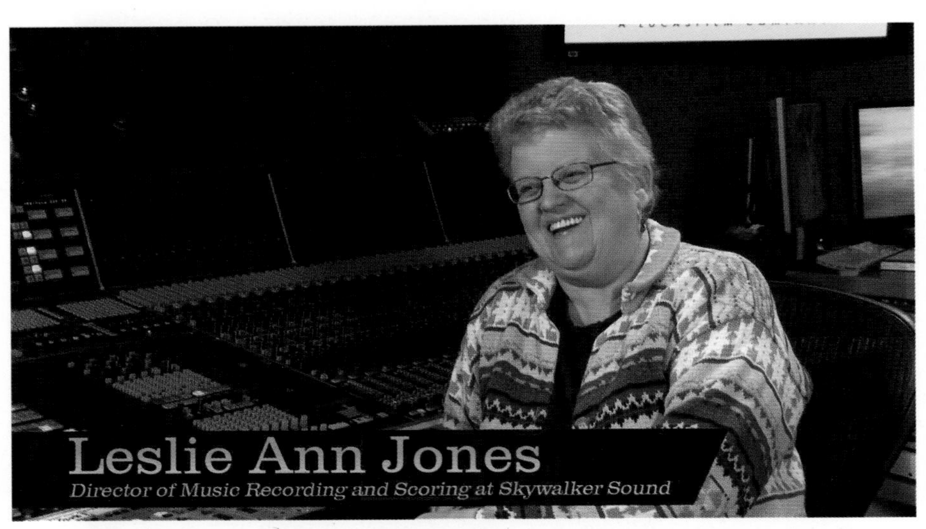

LESLIE ANN JONES: From Sly Stone to Skywalker Ranch. Loved and revered by all.

9. Don't Over-compress the Snare!

We've already mentioned compression for no reason on the stereo bus. It also seemed that, in almost all of the entries, people put compression on the snare. But I heard very few mixes that had a reason for using it. So don't over-compress the snares; just make them do what they're supposed to do. Give me that backbeat!

10. Compete by Standing Out

In these contests, and also in the real-world "Make the Record Company Want to Give You the Job" contest, the people judging hear a lot of stuff. They may be tired, they may not be listening on the best system, and they probably have to listen to many songs in a day. You've got to make yours stand out. Everybody listens to a lot of music. Make yours stand out!

6 Don't Call It a Comeback!

The Hamburger Meeting

HERB: In 2008 Dave and I had gotten back in contact after being apart for a while, and we were spending a lot of time on the phone. RocNation was managing him, so we weren't really discussing business. We were just hanging out as friends, debating what I started calling Dave 3.0. Then one day he came over to my place and Aretha made us burgers. Dave had a very positive reaction to his burger, and somehow that day we got a little more serious about his career and where it was going.

We'd had a similar conversation one day a few years earlier at Larrabee North, where Dave was working at the time. We were both intrigued with what the Web was becoming and it was obvious that there was going to be opportunity. At the time, I was hot as a manager, Dave was hot as a mixer, and I was beginning to wonder if there was something we could do together with the Internet.

DAVE: I remember that first conversation pretty vividly. I recall that I tossed out the idea of a school. True to form, Herb struck the idea down, with the usual enthusiasm with which he strikes down most of my ideas.

HERB: It wasn't that a school was a bad idea! It was just the concept of how the school would be structured.

DAVE: My idea was a 1970s idea and Herb's was more in the realm of 2015.

HERB: At the time of that first conversation, the tools on the Internet had not gotten to where they are now. We tabled the subject, but fast-forward a few years, and we're at that first hamburger discussion. We began to see how the ideas might take shape, and we decided to keep the conversation going.

Since our early days, our friendship has operated with the belief that there were things that we could do together. In between, we'd just check in on each other—in really good times, and in the bad times. The support frame between us is something we've always had, although we never defined it.

The first hamburger meeting was just another example of that. Dave had a great relationship with the RocNation folks. But separate from that, I was trying to examine where he was going to go in his career. Because from my perspective, it seemed that his career had started to shift.

Like a lot of engineers, Dave had never done any real self-marketing. If you wanted to work with him, you had to track him down. You got his e-mail address or his telephone number and reached out. That worked when there were 10 A&R people who were hiring him all the time. But when the business began to go outside of those A&R people, or those A&R people lost their jobs, things changed. And if people don't know how to get to you, and you're not all operating in the same small loop, your career can shift underneath you without you even realizing. I could see that happening, and Dave felt it, too.

Those were the topics at the first hamburger meeting. When we finished that day, we scheduled a second hamburger meeting. But Dave didn't show. That was very unusual. Dave's the kind of person who calls to say, "I'm 34 seconds away!" So to hear nothing was very odd. There I was, with burgers waiting. I called, but there was no answer. That went on for about 10 days. No response. Then, out of the blue, I got the call from Neal Pogue alerting me to what had happened. And I went into tracking mode and into the emotional shock of what had happened to my friend.

It was a very tough time.

RECOVERY

HERB: But let's fast-forward again to when Dave left the hospital and got back home to his house in Agua Dulce. I went out there for a visit, and a few things were immediately apparent. One, he was recovering. Two, he was doing well. And three, he wanted to put a studio in the house! His health and mobility were coming back.

Not long after my first visit to his house, I went with him to a meeting with RocNation and its head, Jay Brown. Jay had known me a long time and was very open to the idea of the three of us working together. And at that meeting, he said something that really caught my ear. He said that one of the main reasons they'd wanted Dave on the RocNation roster was that they knew his name would attract people. That really impressed me. I loved the fact that Dave represented that kind of brand to them. It confirmed what I'd always thought.

After that, I must have gone out to Agua Dulce 30 times. We'd watch basketball together and we'd eat together. We were just in friend/creative mode. But all the time, Dave was really, quite miraculously, getting better.

It was eye opening how consistent and how dedicated he was about his rehabilitation. He was doing all the right things, taking his medications and following the doctors' orders. He was on a mission, and I realized that Dave with a mission is probably the best Dave you're going to get. It's why his mixes are so good—for him, creating great mixes is a mission!

This, of course, was the big mission: his life. His daughter was going to college, he had obligations, and he was approaching recovery like an athlete. He didn't talk much about it. He just did it.

I realized then that as well as I knew him, I'd never actually seen him at his best until that period. It was impressive and inspiring. What we might be going to build together, we didn't know. But the recovery was fully on.

One example of how he showed his determination was that he would take every opportunity to walk with anybody who visited. If you got up and went to your car, Dave would walk with you to your car. If you were going to the kitchen to get a bag of chips, Dave would walk you to your chips. Everybody would say, "It's okay, Dave, I can do it myself."

But he'd just reply, "No." And then he would stroll with you and physically manifest himself.

I'd be remiss if I didn't point out the phenomenal job Dave's wife, Joy, did during this time. Relentless, consistent, and persistent, Joy was a major factor in Dave's recovery. Also key was the power of Dave's love for his daughter, and his daughter for him. Melissa is a gem and she adores her dad. It's a great thing to watch.

Spending all that time together, we returned to our conversations about the Web. It was also during that time that I took Dave on a visit, deep into the 'hood, in South L.A., to the Watts/Willowbrook Boys and Girls Club, where we met with a friend and special human named Les Jones. Les was head of that chapter of the club, a former EVP of a major bank who gave that up to help disadvantaged kids by working in the Boys and Girls Club system.

At that point, I thought it was important for Dave to see the actionable engagement that was happening there, the courageous progress of these folks, and the possibilities regarding education. I thought it would inspire, and it did. Dave and I, and Les and the kids, all left the encounter pumped up, all of us inspired by each other.

PENSADO'S ANGELS

HERB: Meanwhile, all of the ideas that Dave and I were coming up with, of course, required funding. We kept coming back to the question, how do we finance this? Dave was recovering, coming back, but with heavy medical bills, and he'd just spent a lot to build a studio in his house. The thought of him writing checks to fund our ideas was pretty nerve-wracking.

The concept was to take the Dave Pensado brand and put it on the Web. But to do that we needed content, which meant cameras, editing, and more. We knew we didn't want Dave as just a talking head. There was already too much of that around! Instead, we wanted to capture who Dave is and what he does, and to create something larger with it that included education, idea sharing, and Dave's philosophies.

We were aware that, for content capture, if you're going to go beyond your computer webcam, you've got to spend money. But we also realized we'd be throwing money into the ether and hoping it would work. That made for a very nervous time.

That's when the first of what we call the Pensado's Angels came in. Here's how it happened. It was sports night. We were sitting in Dave's living room watching TV, and I said, "I've got to call [my stepdaughter] Tyler." Dave replied, "That reminds me, I need to call Tyler the guitar player." Now, Tyler the guitar player also worked at a digital video network. And when Dave called him, Tyler said, "You should come down and see these video shows we're shooting!"

So we did. We went over with nothing in mind except to meet the folks there and see what they were doing. They had a little in-the-box camera setup, and a couple of switchers, and while we were waiting in this little bullpen area we started talking about our idea to some of the team. Since they were digital guys, I wondered if our idea would seem relevant to them. At first, we were just talking with one person. Then it turned into three, then five, until finally we had a roundtable full of people in the bullpen with us and they were all saying, "This is fabulous!"

The next day I got an e-mail from the company CEO that said, "We would like to do this as a show."

It turned out that part of their raison d'être was to put instructional videos on the Web, and what we were planning most definitely made sense to them. I called Dave, and we met at a sushi place in Encino to talk. It seemed to me that this offer was a way to implement our idea without taking on the cost. If we could work out a deal, we could do it for a month or so, grab enough content to give us a start and enough time to figure out how to monetize it.

Dave agreed, and that became our avenue of opportunity. We structured a deal where we'd keep ownership of the name but gave up rights to the first year's content. It was a great start. But then we had to put in the work that would make the show a success. I went into full-on "Herb, get it right" mode. It completely consumed me. Poor Dave, because we went from idea to whirlwind in a blink. But he was gracious and trusting, even when I asked him to start reading scripts looking at X's taped to his kitchen wall to represent imaginary cameras. He never missed a beat, one of the things I love about him. He's an all-in kind of guy.

Not being broadcasters, we were just feeling our way. But all of the things that have now become staples of the show came about because we were doing all this crazy rehearsing at his house.

Pensado's Place was just a working title, because that was where we were rehearsing—at Dave's house. It was never intended to be the name. "Into the Lair" came about because

Dave likes to mix with the lights down low. I wanted to add a technical segment to the show, but I thought it should have a different name. So I said, "You know, you kind of mix in a lair, so let's call this 'Into the Lair' for now. Which, by the way, Dave hated.

DAVE: Hated might be a little strong!

HERB: "Corner Office" was originally a way to bring executives into a segment. The interview was another segment. We were just messing around with various ideas, and then we moved to the physical production of it all—still sitting around his kitchen table. We put marks on the wall, I wrote up some rough scripts, and we started practicing.

It was so bad, it was horrifying. Dave is a team guy, so if you put him inside a box that is uncomfortable for him he'll give you a couple of tries. In this case, discomfort turned into frustration pretty quickly, and after two or three of our rehearsals not going well, we had to figure out something else. In retrospect, that's another funny thing, because what that process led to was the on-air gig I have now.

I'd started feeling that it wasn't fair to put all the weight on Dave to read scripts that I'd written and also to basically carry the show. So I decided to find him a cohost, figuring that somewhere between the cohost and Dave, the dynamic would sort itself out. But then Joy, Dave's wife, called and said, "I've got a great idea! You should cohost."

I was so not into it. "No, no, no. I'm a behind-the-scenes guy!" But Joy insisted, saying, "Dave respects you. You're the only person who can direct him, the only person who can cut him off when he needs it. He's not going to pay attention to someone else who tries to cohost, and that's just going to give you a bigger problem!"

All of this, of course, coincided with the network being ready for us to do the pilots. They were set to go and I didn't want to hold things up. And although I thought we were terrible, since we'd done all those rehearsals together, Dave and I had developed this weird sort of rhythm. It wasn't great, but it was something. So finally, to keep things moving, I agreed to be the stand-in cohost for the pilots.

The network, which at the time had about 22 shows on the Web, had us do two pilots, back to back. And they loved them! So we thought, "That's cool." Because it was still a limited idea: we'll capture content, we won't have to pay for it, we'll get started.

Then I got an e-mail from the network saying, "By the way, you've got to cohost." My response was, "This is just not gonna work. I've got a full-time gig at Kalimba. Plus, I'm a behind-the-scenes guy and I have to manage things. No way. I can't do this." But in the end, I didn't want to leave Dave hanging. With the pilots done, they wanted us ready to go in two weeks. We didn't know how long the deal would last, and we didn't have time to find somebody else. So I gave up, jumped in, and went.

So there we were, doing a show. And before we knew it, we were into something that required us, every week, to be in a pattern, and to be prepared. We had a job!

Meanwhile, Dave was still mixing and I was still at Kalimba. A couple of months in, we began talking about having me manage him again. We were working together daily anyway, and it started to look like it might make sense. So we had another meeting at the sushi place and talked it out. Managing Dave was something I could do as a friend, on the side. Plus, he needed it.

Dave's got a big heart, and although he's very direct when he's dealing with philosophical issues, he's not always very direct when dealing with business. So it can happen that sometimes people who are friends of his—not that they're trying to take advantage, they just want a mix by the best, and Dave wants to help them. But when I looked at his portfolio of things to do, there were a lot of people who weren't paying, all with demands on his time, with Dave not wanting to be direct about it. So I started managing the engineer side and cleaning things up. And there we were: the Dave and Herb show again.

The unintended consequence was that, for the first two months, I was so busy that I didn't get around to looking at the reactions to the show and the comments that were getting posted. Joy looked at all of them. She'd tell Dave and he'd tell me, and I was thinking, "Aw, that's cute!" But when I started paying attention, that changed to "Wait up! Something is happening here!"

WE'RE CATCHING ON!

The first clear sign was the recording schools. There were six schools that wanted to work with us; then there were 12, then 15. The other phenomenon was that the comments were all so overwhelmingly positive. It wasn't casual. It was more, "I've been looking for this forever; Dave is a genius! Where has this been? This is my YouTube porn!"

The reaction was coming from all over, not just from Los Angeles and the other music centers. It was starting to look like there was something to build on. Here's the other thing: Dave and I started to like doing it! Then, because Dave had this golden address book, the first 20 or 30 guests were insane, top-triple-grade-A people. And *they* were all saying, "This is cool as shit, Dave. This is really great!"

At the same time it was also pretty uncomfortable for me. I'm not an engineer or an audio person, and here I was cohosting an audio show. I was trying to be an on-air show runner, trying to avoid getting in the way of the information, and at the same time trying to stay connected. I was learning on the spot, in front of people—and I was fascinated, because we had smart, cool people on every week. It was building. Before long, we saw we had something that needed protection.

So we went for it, full on. I went into hyper-psychotic mode: "Get it right. Get it right. Get it right." Dave was a booking machine every week and we were getting immediate feedback through the Internet. We had the machinery of the network to take care of all the stuff we knew nothing about, like posting, annotating, and algorithms.

Then Pensado's Angels intervened again, with our sound guy. In the beginning, the show had various different producers. They ranged from fashion models to bearded literature geeks, none of whom had anything to do with audio. They were just putting out content. Our sound guy, however, was consistently the same. He would come in after each show and sit with us. He was very savvy about things, and Dave and I figured out that he was really kind of producing the show. His name was Will Thompson. Pretty quickly, Will became the actual producer of the show. Now we had someone who was trained as an audio engineer, who was also great as a digital publisher, and who was wickedly, ridiculously smart.

Will and I started spending enormous amounts of time on the phone. I saw the potential from the point of view of how the product could be good for the audience. Will saw the potential and understood the digital execution standpoint. That combination of elements proved very effective. In the beginning, it was critical. Will was also a psychotic, all-in worker with multiple hats—a blood brother and perfect fit.

From my viewpoint as a friend, Dave was now flourishing. He was back on track as a mixer, his clientele was picking up, his studio was smoking, his mixes were good. He was a guy returning from a place that was not that great to be in, and all these other things were now imbuing him at the same time. His work was growing, we were getting his business straightened up . . . and we had this show.

We used to meet each Wednesday in the back of El Pollo Loco in Sherman Oaks and ride to the studio together, the whole time saying, "Can you believe this shit?" On the ride back, we'd review everything. "Did we do this well? Did we do that well?" Almost without realizing it, we were becoming committed to this thing.

IF YOU WANT TO DO SOMETHING, MAKE IT AS GOOD AS YOU POSSIBLY CAN

HERB: Once we'd gotten the management thing going, we had a process. We had a show to shoot every week, we had to prepare for it every week, and we also had mixes to do. I had to handle the business related to the mixes. Dave had the pressure of finding a guest. We weren't yet thinking about guest strategy, genres and such. We just had an hour to put up every week. We didn't know if it was ever going to make any money. But at least we weren't coming out of pocket.

We hadn't started out with any other goals, financially. But before we knew it, we got a little advertiser. That was interesting! Next, we were approached by Vintage King. We went to dinner and they said they wanted to sign on. We walked away from dinner going, "Oh, hell, they're going to sign on for a year. Now we have to do this for a year! Shit."

Then Avid reached out; they thought what we were doing was cool. Before we knew it, we were committed for two years because we had advertisers. Once that happened, we got serious about the fact that we were really doing TV. It didn't matter that it was just on the Web. We were going to approach this from a production standpoint like it was television, to the degree that we could with the tools that we had.

So, from how the two of us were on air, to the look, expectations, graphics, transitions, all of that, we had a particular standard. We didn't want to be just another webinar, webisode, tutorial kind of thing. The production elements and the music had to be high quality. Of course, at the time, we didn't really know who our audience was going to be. Were we talking to people who were all Web based, so should we just be speaking to them in Web language? We weren't sure. But our guests were, and are, not 16-year-olds. They are 18-to-40-year-olds who grew up on TV. So if someone was going to consume us, to me it seemed that they needed to consume us in a way that was second nature to them. Especially if we were presenting ideas that were a little obscure or even disruptive, the delivery of those ideas had to work seamlessly for the audience. That meant TV, to the extent that we could manage. And so I became pretty much a stickler, or, some might say, an asshole, about that.

DAVE: Herb had high quality standards for the show, which he's also had for every element of his entire career. So it wasn't really a shock that he'd want this show to be as good as or better than anything that's out there. He wouldn't allow limitations because of our situation. He was focused on the potential of what we could do. From the very beginning he emphasized that we had to make good TV. I heard those words a lot. Even when we were practicing in the kitchen, it had to be good TV.

Here's something interesting about partnerships. If two partners come to a relationship with identical goals, skill sets, and personalities, it's not a partnership. It's just a doubling of effort, which doesn't really help.

My motives and goals and what I brought to the relationship, and to the show, were very different from Herb's. As a matter of fact, at the beginning, I thought we were doing something entirely different. What happened with the show was that it wasn't my ideas that got executed—it was my ideas filtered through Herb's vision.

HERB: And we've ended up with the school that you originally wanted!

DAVE: But the reciprocal of that is also true. Herb has been passionate about preserving the concept that the show would be something that could bring this wonderful thing called audio to the world. And I mean to *the world.*

He had all these ideas like the branding, and helping me out with my career, but they were all predicated on the concept of free education.

HERB: It's true that was the impetus for everything. Dave has envisioned a school of some sort for 20 years. I always knew, at the root of it, that one of the best Daves is the teacher/philosopher king, and I kept looking for how to construct the parts so that we could combine his love of teaching and audio and philosophy with the salon vibe into a vehicle that could go out into the world.

In the show's first year—other than making sure that people got educated—Dave didn't care so much about form, how to present, all that stuff. But how to present was always a big issue for me. It also became important to him a bit later when he realized the impact of doing it well was that you got better time with your guests, you reached more people, and you got more done. But teaching and education were always at the root, uppermost in Dave's head. Consequently, when schools started adopting us, my reaction was, "Whoa, we really are onto something." We just got there much quicker than I ever could have imagined.

DAVE: I never thought of Herb as having less than equal passion for this endeavor. And his passion for doing this was not a fun thing on some days! But it carried us through. His passion for this process and for trying to bring this to fruition was no less than mine, but it was just slightly different from mine, and that's why our partnership works.

We're very similar in a lot of ways and we're completely different in a lot of other ways. What binds us is that we tend to find a common goal, which means the post facto tends to be pretty spectacular.

Also, we both have athletic backgrounds, so we're both pretty competitive. Bottom line, the show would not have happened without Herb. Herb could have done something without me. I couldn't have done something without Herb. I'm just not constructed that way.

BRINGING AUDIO TO THE WORLD

DAVE: Let me give you a bit of background. Part of what I do is altruistic, and part of what I do in terms of sharing is ego. I have an ego, a great big ego. So does Herb. But we complement each other's egos. Part of it is going out with your buddies and catching a bigger fish. Which is kind of like the show.

It's more sharing than teaching. I don't consider myself a teacher. I just share some of the things I've learned along the way. When you have a passion for something, as I have a passion for audio, sharing that information is just somewhat normal and expected. It's like in athletics. The veterans don't exactly teach the new kids coming up, but they share information that's going to help them move along in their profession. Just like they had information shared with them by the people before them.

When I first got into audio I was so naïve, I didn't even know schools were an option for learning about recording. There weren't many of them at the time. But a handful of people did for me what Herb and I are doing through *Pensado's Place*. We're just doing it on a broader scale.

Initially, the show wasn't about the big picture of audio and all its possibilities. I chose the word "audio" because Herb felt that a focus on "mixing" would be too narrow. I always trust Herb, but at the time I didn't see how that was the case. I wanted the show to be about mixing. But then we had Alan Meyerson on the show, one of my favorite engineers, as you have heard previously. We had him on the show as a film mixer and it was one of our best shows.

That was a turning point for me. I always supported the concept of broadening the show beyond mixing into the audio space, but wasn't sure how to do it. Then Herb came up with the idea to go to the stage where they shoot *Let's Make a Deal*. And I'm thinking, "Aw, c'mon, man. I'm an intellectual. I don't even want to watch *Let's Make a Deal*, much less care about having anybody involved with it on our show!"

But it turns out Cat Gray works on *Let's Make a Deal*, and I have so much respect for him. I learned so much on that episode that it turned out to be one of my favorites!

HERB: Cat was the keyboard guy for Sheila E and worked with Prince during the *Purple Rain* days. On *Let's Make a Deal*, Cat plays all the music live! I have been blessed to work with and to be around incredibly talented musical artists—Stevie, Whitney, David Foster, Pharrell, Quincy, Dre, Celine, Aretha, Michael, and on and on. Cat is easily in my top five of baddest musicians. I felt strongly that Cat's musical backstory, combined with the only game show that uses live musical improvisation, made for a great show, and it worked.

Check this: two days ago Cat and I got on the phone with a 13-year-old who saw the show and dressed up as Cat Gray for Halloween! His uncle had contacted us to tell me. How can you not love being able to create heroes for people?

Another focal point for change during the first year of the show was the Olympics. We'd already started pushing the audio envelope. I wanted movie people, Foley, gaming, all different kinds of audio. Then I read an article in the *New York Times* that described what went into capturing the sound for the Olympics. And we had Dylan Dresdow on, who had worked on the Super Bowl halftime show with Black Eyed Peas. Hearing about how all that worked started me looking even farther afield for different scenarios where we could highlight audio people. There are so many fascinating things out there going on in audio.

For something as big as the Super Bowl, you have to serve a multitude of masters: the television production company, the act, the label, the halftime committee . . . there are all sorts of people who have to sign off. Then how do you capture the complexity on the field? What are the microphones? How do you make sure that audio quality gets broadcast out to viewers in sync? How do the Super Bowl and the NFL plan audio for this event, and what are the numbers of technical people involved? I'm excited about showing all that kind of detail.

More and more we started imagining things, we would dare to ask, and people would say yes. We got cleared by major L.A. film studios to shoot on their lots—Universal, Sony, and Fox—because Ward Hake, a VP of music for 20th Century Fox, was a fan and was a guest on the show. All of this gave us more impetus to continue to push. Creating a paradigm shift is a lonely pursuit. You may be surrounded by doubters and you must have huge cojones to continue. It's risky, particularly at our age! But when it works, it's super gratifying. Life really is for living.

IT'S REALLY HAPPENING!

HERB: Then one day someone recognized us when we were together. Dave and I had just come from taping the show, and we were eating in a chicken place in Encino. All freshly scrubbed of makeup, we're having our chicken and a guy comes up to us and says, "Oh, my gosh, it's Dave and Herb from *Pensado's Place*!"

After he left, we looked at each other and said, "Are you kidding me?"

We accepted that we had something. And when we realized the opportunity, and began wondering what we were going to do with it, we also realized the responsibility. There were now people who were dependent on us. Many took it way past audio education to something personal, with all the consequent human interaction. That became heavy. The final part was, early on, we had no idea people would find us to be entertaining! We hadn't planned the entertainment part.

I think, for both of us, what really gets us is that the show is seen by intelligentsia as about intelligentsia. We hit the bar for really critical people, and we hear from schools all around the globe who tell us, "We teach with your show. We think it's intellectual! Plus, you make it fun and interesting."

SHINING A LIGHT ON THE PROFESSION

DAVE: An interesting thing about partnering with Herb is that the original idea may be minuscule, but then it produces tentacles and branches, to mix a metaphor, and grows into something that you couldn't have predicted.

Along the way, this passion and love that I have for engineering and mixing has broadened. Back when the show started, there was some cloudiness over where the profession of audio engineering was going. The landscape kind of looked like a Discovery Channel show on the drying up of the Serengeti lakes, with all the fish now stuck in a tiny pond with their fins flopping around.

With that picture in my head, it seemed like this responsibility that we had been given should be utilized to show people that you don't have to be a mix engineer, you don't have to be just a classic recording engineer, you can go into Foley, or television, or so many other areas. That's when I fully embraced Herb's concept of broadening the scope of the show.

HERB: And to help confirm that direction for us, people responded positively to it.

DAVE: I responded positively, and I tend to be a typical viewer! In some ways—and I know this metaphor will piss off a few people, but those people are not engineers, so it doesn't matter—I look at the show like when I was a little kid and I used to grab a magnifying glass and focus the sun's rays on an ant and just torture the little guy. As a small child, I didn't understand exactly what I was doing, but now I'm doing something similar in a way that's beneficial. We're not the sun; we're the magnifying glass that focuses the sun in such a way that it can have an impact on people who view the show.

Now, hopefully we don't scorch our audience! But, as Herb and a lot of the engineers I respect have pointed out to me, some rather emotionally, that's one of the things they love about the show: it has brought a focus to this process of audio engineering that they really like. They like the fact that people recognize them on the streets. The show, in a weird metaphoric way, is not ours anymore. It belongs to the audio community. We just kind of hold the magnifying glass with full cognizance of the responsibilities that we have.

HERB: We specifically treat the show as though it's not ours; it's the audience's. And in many ways they power us. They write in, subscribe, comment, etc. Some suggestions we take, some we don't. We stay open to their ideas, and they communicate with us weekly.

We're showing that there are other jobs besides classic recording and mixing, and we're helping to bring attention to the overall world of audio. But there's something else: we're redefining what the audio world today actually is.

Everybody has tools in his or her computer to be in the audio world. It doesn't have to be the way you make your living, but it can still be your passion. A huge part of our audience is made up of people who aren't interested in being audio engineers; they're just plain interested in the show.

If you take the number of people who can use GarageBand or some other program and enhance the videos they post on

YouTube musically or sonically, it's a much bigger audience than those guys who are trying to become Dave. And those people find value in the show. So the more we're passionate about the space, and about expanding the space, the more we pick up people who are just interested in audio itself.

That mission is to show that there's this audio world and it's got layers to it. You didn't know about it before, but it's really interesting and it's got smart people in it—and your world doesn't exist without it!

DAVE: Cable TV changed a lot, because it opened the door for so many things that now need to be seen. Initially, no one thought that you could sustain multiple cable channels by having programs about things like cooking. People didn't particularly want to become chefs, but the Cooking Channel is now huge because everybody wants to have the same tools the chefs do. And the level of culinary expertise in the world has truly gone up—not because everybody wanted to be a chef, but because they liked watching the chefs. The concept was created: "Wow! There's a market for shows about cooking!" Then there were shows about photography, and antiques, and how to decorate your house.

So when what Herb calls our Angels put us in this space, we encountered all this passion we hadn't known was there. It's now pretty clear that people want to take better photographs, cook better meals, and decorate their own homes. They also want to know more about audio.

HERB: The strategic part is that cable and the Internet, particularly YouTube, made niche important. Once niche became important, you could go into a specific area and focus on it and find an audience. We started recognizing that we could go down that route. We just didn't realize how big it could be.

If you analyze all of these successful niches you see that, invariably, they have to grow stars inside of them to keep that niche growing. I can name every star on the Food Network, because I watch it. Discovery has a set of stars. They need that in order to keep it going. When you look inside, none of it is accidental.

In the audio space they've hooked onto *Pensado's Place*. When I was recognizing that early on, it was driving Dave pretty crazy, because I was getting intense. "We've got to do things a certain way. We can't be sloppy here; people have patterns and expectations and we have responsibility." Once we embraced that, we saw good reactions really quickly.

We're talking about the comeback trail here. And seven months in, when we saw ourselves going from Dave using a walker to producing an audio education concert—our first offstage show where people paid 100 bucks and flew in from 11 countries around the world—well, we were stunned. They were lined up around the block on Sunset Boulevard and we were floored.

From wheelchair, to walker, to that event is less than two years. What I tell everybody about *Pensado's Place* is that it is really a story of redemption, an incredible comeback from a health crisis. And that if you try in your life, who knows what can happen? We are two old dudes on YouTube for an hour. We're doing nothing the way we should! We're not young, hip, and clever.

We have a certain amount of expertise, we have a certain amount of intelligence, and yes, you can research and find the tools you need to produce a show. But it's different when you have to get up and do it each week. When the camera comes on, you've got to deal with it. And you're dealing in front of people, and they get to comment. I didn't want to do the show. I'm thin-skinned! I was sure they were going to flame me. But I couldn't let my boy down. And if he's going to get up there, coming from where he'd been, what was I gonna say?

DAVE: Well, we wouldn't read the comments at first. My wife read them to me.

HERB: But when we saw they were good, then, for about four months, I turned into a crack addict about them.

DAVE: It's what drives us.

HERB: I wake up every day and get this love that is so intense and I realize, these are not stupid people; this is not hero worship. There is some value they are finding that for them is very personal. Also, it has always been more about passion than audience size. We were smart tactically about that. We went for audience engagement versus views. Remember, we have to work harder for our engagement level to be so high for a whole hour, and for people to be so passionate about it. But that's what's made us.

DAVE: It's extremely rare that the number of views on YouTube is accurate. They are always massaged and manipulated in one form or another, legally and illegally. But engagement can't be manipulated, and that's where we shine. We shine because of our audience. They really have taught us well.

HERB: So, to put a button on it, in terms of a comeback, I think it is fair to say that it's about the show first, and everything else—all our other work—after that. The show is core to what we do. It's not that mixing is any less important to Dave, or that any of the other things I do are less important to me. But getting the show right every week is paramount. Unlike most shows on legacy television, we don't have a hiatus. We are on every single week. If we take time off, we've got to have a program to fill that week. It is 52 weeks of making sure the audience isn't disappointed.

DAVE: I want to throw my little button in there, too. Whenever I attempt to manipulate and take control of my life and career by trying to do things that will bring me money, fame, or fortune, it's always been a disaster. So early on, in my 20s, I gave up trying to control that. Instead, I tried to surround myself with good people and to make decisions that were just decisions—not categorized as good or bad. I also noticed early on that whenever I tried to do something to help someone else, I was rewarded for that and blessed by that. I really believe that, with the show, Herb and I have been so desirous to put the show first that the blessings will come.

HERB: This whole comeback has been about us being who we already are, managing through whatever life has thrown at us. And when it became "We'll put a fish eye on you guys," we said, "Well, then, let's do something in front of that fish eye." And we figure it out.

I get to work with one of my best friends, and we have people we are positively affecting in ways that go way outside of *Pensado's Place*. What drives us is the responsibility that we know every week we have to deliver. The gift is amazing. But I wouldn't say the comeback is complete.

DAVE: The comeback is kinda complex, because Herb's much more than management. Sure, Herb's a manager, I'm a mixer—and the show has allowed me to be a better mixer and have more fun mixing and for him to be a better manager and have more fun managing! But now it is so much more as well. Our worlds have expanded. We're on-air talent, and Herb's a producer and a show runner, too. We're putting on live events. And we're still mixing and managing! But to me, the single most important thing that took this from a good idea to a great idea was Herb's insistence on quality and doing things right.

INTO THE LAIR, NUMBER 73: CREATING VINTAGE VOCAL DELAY EFX TRICKS

DAVE: I was trying to put together a cool idea for an ITL and I started thinking about some old school techniques that we could do inside Pro Tools. I first spent a little time creating an old school "ducking" example for you. Then I started thinking about what the differences are between ducking and side-chaining. Back in the day, I made my own compressor because I couldn't afford one. I realized that it was just an internal side chain, and I created a ducking technique.

Back then, you'd put the guitars through a compressor, then you'd put the vocals through the side chain. And when the vocals came in, the guitars would drop down automatically. That's still used a lot, by the way. It's a great technique in the analog world. There are a couple of places you can find a side chain—the Drawmer gate has one, and EchoFarm has a ducking function. But I digress.

What I'm going to show you here is a way to get feedback on a delay, old school style. Now I know you're thinking that I've got a perfectly good, working feedback button; why don't just I slide that feedback button over to get something I want? Well, you could do that. But let's think about what feedback does. It takes your delay and it sends a piece of the delay back into itself. Then that piece comes back out of the delay and gets fed back into itself again. That's why it's feeding back. It's a loop, and that's how you get the delays. Also, each time the feedback goes through the loop it decreases in volume.

Now, I'm sitting there thinking, well, that's kind of boring. But what if I put something before the delay, so that every time the feedback goes through, something new happens? So I decided to take a bit crusher, like Lo-Fi, or it could be any effect you want—for example, it could be a high- or low-pass filter. What happens is, every time the signal goes back through the delay on the feedback cycle, it gets more and more and more of whatever effect you've chosen.

So I take my vocal, which is coming off a track, and I just roll a little of the low end off, and the track sounds pretty good on its own. But then, for the delay, I send it out to two buses, 61 and 62, using two mono auxes, because I want ping-ponging. On 61 I've got a quarter-note delay with no feedback. On 62 I've got a half-note delay with no feedback.

You only hear two delays: first one left, then one right. Now pay attention. The left delay is being fed from bus 61.

I'm going to send a piece of the right delay back to the left delay, and that gives us a repeat. Let's do the same thing the other way—send a piece of the left delay to the right delay—and we've created our own feedback loop!

Now for the fun part. Every time the signal goes back through it's degraded a little bit. I'm using Lo-Fi, but you can use anything you want to effect it and you can set it however you want. I've got it set up so the delay is going to stay the same but it degrades a little every time.

Experiment with that. You could try a real narrow bandwidth, or you could automate the bandwidth so each time the delay recycles back through it will have a different tone. You could automate a pitch shifter and each time the delay comes through it could be a different note. There are all kinds of techniques you can use, once you have access to that feedback loop!

7 Dave the Philosopher

The Rule of Thirds

DAVE: In my opinion, looking at music as if it's separate from painting, photography, theater, etc., is probably not the best way to go. Instead, it can be helpful to think about music as being a part of the rest of the arts. Often, I'll find that things that may confuse me about music don't confuse me as much if I think about them in relation to the visual arts.

Of course, you can argue that rock and roll and pop music don't have a lot of art in them. But in fact, art is composed of both art and craft. Some of the music I like is mostly craft and very little art, but that doesn't mean you can't apply the principles of the art world to it.

In terms of assimilating art, there are certain things that the human condition tends to feel comfortable with. One of those basic things is the rule of thirds. For example, we don't like to see a photograph with the subject dead center in the frame. It may not give us a totally negative feeling, but it does make us feel a little uncomfortable. That's because, as it turns out, we tend to like our subject matter in photographs and paintings on a grid of thirds.

To demonstrate, if you divide a painting vertically into three equal thirds, and then horizontally into three equal thirds, you end up with four nodes where the divisions intersect. And if we're taking a photograph of a friend, one of those nodes tends to be where we'd like our friend's eyes to be—preferably the higher node. With landscapes, we like our clouds and our horizon along one of the horizontal lines. Sometimes you want more sky, sometimes less, but we tend to find it a bit unsettling for a photograph that's on an 8-by-10-inch page to have the sky a half-inch from the top. Although a lot of great photographs actually do have that!

Before we get too far in and the reader starts thinking that this sounds crazy, we should note that the techniques and philosophical components of the art world are as pleasing when they're broken in an artistic way as they are when they're implemented in an artistic way. When you look at the work of some of the greats, you'll notice that these techniques can be manipulated in such a way that they'll create a whole new experience.

In my profession of mixing, I like to think about constructing a mix visually. And because my personal visual references have all come from playing music on stage in live performance situations, I like to construct the stage.

Some mixers like to construct a recording of a live band from the audience's perspective; some prefer to use the performer's perspective. If you put the hi-hat on the right, you're coming from the audience viewpoint. If it's on the left, you're coming from the performer's side. I like the performer perspective, so, since most drummers are right-handed, I pan my hi-hat track to the left, and my shakers and tambourines to the right. Once you've made that decision, you're bound by a musical version of the rule of thirds.

It's not a direct one-for-one correlation, of course. But, fundamentally, you're creating what you see visually in the audio space. You pretty much want to end up with something that gives you the same pleasant feeling that you get from seeing something placed on one of the nodes where the thirds of an image intersect.

Of course, everyone will interpret this a little differently. Personally, I like to pan based on the fact that, in the Western world, anyway, we read left to right. So I like to resolve things left to right. That could also be perceived as a part of the rule where one node divided into three is in the far left position, one's the center, and one's the right. We call that LCR (Left, Center, Right) in the stereo field. Thus, LCR is, in a way, based on the rule of thirds.

Those sections are the most sacred spots in the mix, the places where you should have a really good reason to put something—particularly the center, of course. In popular music, the kind I'm generally working on, we like the lead vocal in the center. Then, if we've got background vocals, we have to decide: Do we want them panned hard left and hard right? Or do we want them brought in a little bit? Or do we want them in the middle? Well, we probably don't want them in the middle, because, as I said, that's sacred territory.

The sound field has shifted a bit in different eras. In the early '90s, we often liked background vocals to be even wider than the speakers, so we'd use techniques to get them wide. Now that doesn't sound modern to us. We tend to like to hear them a little closer in. If you're reading this book in 2050 we may be 30 percent outside the speaker field, or we may be mono again!

How we treat the space is one of the elements that suggests whether a mix is contemporary or not. In the rock world, guitars are hugely important. But you still want the drums and the kick and the snare in the middle. So what do you do with the guitars? If you put them on the edges, it's most likely going to feel a little disconcerting, because it doesn't feel like a live performance. So probably you'll want to bring them in a little bit.

Now, if you're working with the rule of thirds, there's no actual space between center and the edges—so where do you put those guitars? Well, you can create a new third by dividing the exterior thirds into thirds again. Then you can put your shakers, tambourines, and guitars on those thirds within the thirds. And that seems to be pleasing to us. Now, if that's not pleasing to you, then create your own rule of thirds! But it's very useful to have a concept in mind, something that helps to guide you and gives you a sense of balance.

One of the things I notice about rough mixes that get sent to me is that my head wants to tilt to one side or the other to compensate for the fact that there is too much stuff on the right. But recently, on the show, I questioned one of our guests about the fact that he had so much happening on the right. In this particular case, it wasn't a negative thing—it worked! That could be conceived as a manipulation of the rule of thirds in a way that was probably not technically the best. But it worked because it drew my focus to parts that I wouldn't have noticed as much if the sounds had been split equally across the stereo spectrum.

The point is, the rule of thirds is an example of how borrowing techniques from another medium or art form and applying them in our audio world can be very useful. It's something that can lead you into your own new ideas, which are ultimately what you want to achieve and market.

We don't need another tube of toothpaste. We've got every version possible. If you're going to come up with a new form of toothpaste, you'd better make sure there's a need. Once you get that need addressed, and people know the brand name, then you can expand. But rather than trying to be the world's most average engineer, doing the same thing as everybody else, I recommend that you pick a specialty, an area where you can excel, where your natural tastes become valuable, and then expand from there.

THE GOLDEN RECTANGLE

Another application of the rule of thirds is the golden rectangle. The most common example of the golden rectangle is a classic chest of drawers, where the top drawer is the smallest, the next drawer is a little taller, the next drawer is a little taller still, and the bottom drawer, the fourth drawer, is the tallest. The ratio by which those drawers increase in size is a mathematical formula called the golden rectangle, and we tend to like it.

The golden rectangle can also be manipulated as a spiral. If you cut a nautilus shell in half (go online and check this out!) the chambers that make up the spiral increase in size as the spiral expands, with those same mathematics that are pleasing to us. It is a concept that works in nature so thoroughly that it gets repeated over and over again in numerous life forms.

I used this concept a while back on a couple of songs. I was thinking that, instead of having a delay repeat in eighth-note increments (where you'd have the original, then, an eighth note later, a repeat, etc., etc.), what if you repeated based on the golden rectangle? So the distance between the original sound and the first repeat would be the top drawer, the second-drawer repeat would be a little wider, then on and on.

I found the best way to do that was to copy and paste the delay. I haven't done the math—I probably should! But instead I just do it by ear. I could sing it to you, but you wouldn't be able to hear it in this book.

DIRECTING THE EAR

Something else you'll notice in the art world is that a great painter or photographer will direct your eye around an image. They don't just let you look at it and absorb it as one rectangle or square. A really great example of this, called *Las Meninas,* was painted in 1656 by Diego Velázquez. The way the painting itself directs your eye is spectacular. It does it in a counterclockwise motion, which is interesting since we would think that clockwise, which tends to be the most natural way for most of us to look, would tend to be in line with the rule of thirds.

You see Velázquez himself at a big easel. Then your eye is immediately drawn down to all of these strange characters that are part of Spanish royalty at the time. Your eye goes up and around them, almost like the spiral of a nautilus, until eventually you see a mirror with a fuzzy reflection and you finally realize that you're the king or queen, and the mirror is reflecting you looking together at this scene. That's the way a good mix should work, like revealing the layers of an onion.

COMMERCIAL VALUE AND VAN GOGH

These kinds of techniques all point to how we can elevate the craft of what we do to the point where it has commercial value. That's what craft is about. But it's also okay to inject art into that craft—always remembering, of course, that in the world of commercial music, we're generally working with paper plates as opposed to fine china. You'll generally make a classical or a jazz record differently than you will a pop, hip-hop, R&B, or rock record.

Commercial value depends on people liking what you do, and, to a certain extent, also liking you. But at the same time you should realize that when you please everybody, you probably haven't really accomplished very much. You've just kind of defined what mediocrity is about. Now, I know I said earlier that the more people who initially like your mix, the better your chances are of having a hit. That is true, but there's more to this concept.

In a letter that he wrote to his brother Theo, Vincent Van Gogh said, "One should never trust the time that one is without difficulties." That made me think about his history. Vincent was so sad for so long, and the only painting he sold in his lifetime was to his brother. That's dedication! That's the epitome of commitment to your concept—even though people may not like it. Now his paintings sell for hundreds of millions.

With Van Gogh's thought in mind, I came up with the idea that if you play your work for 10 people, and all 10 of them really like it, you've really not accomplished anything beyond creating something average.

I think instead a good result to try and shoot for is the 3/3/4 ratio: if you play your work for 10 people, 3 should violently hate it, 3 should say, "Not bad, Dave. I like it," and 4 should say, "Whatever it takes, I want this! Whatever I've got to do to get this, I will do. Money is no object. I've got to have it!"

Of course, this is a formula that's hard for me to implement all of the time, because I need to be liked. But at the same time, I believe that when I acquiesce to just being popular, I don't do my best work. I'm at my best when I try to go for it and push the edges. Now, when you're near the edge you're going to fall off a few times. But that's okay.

Here's my best interpretation of this concept: Don't judge your success by whether you achieve universal acceptance of what you do. Judge your success equally by your failures. They are as important as your successes. Meaning, while having all failures is probably not a good sign, if you have only successes you may not actually be achieving your true potential.

In the history of popular music, you'll find that extremely successful artists often, and rather quickly, turn into unsuccessful artists. One of the keys to both success and longevity is the 3/3/4 formula.

Taking chances that may fail is especially important when you're starting out. Think about it. When you're starting out, do you really want to compare yourself with the already greats? They have a big head start on you. Instead, why not use your best self as the standard of comparison? Because it is possible to be the best "you" that you can be. And you can be unique. If you stick to emulating what the greats do, you're always going to be a third-rate copy of them. Better to be a first-rate original of yourself than to be a third-rate copy of the top engineers, the top painters, or the top chefs.

In order to be that first-rate version of yourself, you have to be comfortable with having some failures along the way. Not all failures are good failures, but most of them can be, because you should be able to learn something from them. In order to determine what's a good failure and what isn't, a network of people you trust can come in handy. It's good to have people around who can honestly tell you if what you're doing just plain sucks or if, although it's not perfect, it actually has to do with growth.

I'm blessed to have assistants and a manager who care about me, so I tend to get good advice, like "This is good, Dave. People are just not feeling it because you went a little too far. Dial it back and they'll like it."

OTHER SOURCES OF INSPIRATION

I said earlier that I don't like to mix genres. But I do like to listen to all genres and inject little pieces of them into each other. Something I've loved for years is going into the dance world, picking little things, and injecting them into the R&B and hip-hop world. If you listen to one of my early records, "Get the Party Started" by Pink, it's got a lot of dance music—which is what we called it at the time, not EDM—elements in it. You'll hear a lot of little techniques that were originally done in the dance world that I thought were appropriate for pop.

So while I don't like actually combining genres, I like mixing them up a little bit. There's nothing wrong with a rock guitar in a hip-hop song or hip-hop drums in a rock song. But when that injection gets to be too much, it bothers me, because it can dilute what's cool about each particular genre. You have to be careful about that.

You don't want it to be like you're sitting down to eat a meal, then taking your fork and stirring all the food together. It's difficult to taste the individual elements—the potatoes and the green beans and the turkey—when they're all mixed up together like baby food. That won't taste like anything! It's important, when you combine elements from genres, to ensure that you can still taste the individual flavors and that they don't get diluted and become baby food mush. You don't want baby food peas and carrots—where I defy you to taste either peas or carrots!

It's okay to mix things. Just don't do it to the point that the mix doesn't taste like anything at all.

WHAT PEOPLE PAY ME FOR

Another philosophy that has served me well is my understanding that what I'm paid to do is not what most people might think it is. For example, I don't get paid for my ability to EQ vocals. I get paid for my ability to take a vocal and manipulate it in such a way that it moves millions of people. But the person who has to be moved first is me. So in reality, my taste is what people pay me for. I think that's a universal concept in any creative endeavor; your taste is what you get paid for. So while it's true that my profession of mixing records is somewhat specialized, I think if we look at it in a broader context you can see that, generally speaking, uniqueness in your taste is something that's worth striving for.

The next comment from the reader might be "That's wonderful, Dave, but how do I get taste?" Well, the way you get taste is by picking the right parents, picking the right environment to grow up in, choosing the right friends, and then understanding at some point as you mature how those parents, that place where you grew up, and your life experiences have shaped you into who you are.

You are a product of your experiences. They are the soil that your taste grows up in. Without life experiences, it's hard to have that fertile ground for taste to grow in. And without taste, you have nothing to sell. You're just another person EQ'ing kick drums. Since there's a jillion of those people a month coming out of schools, your taste is probably the most important thing you have.

The second part of this is the ability to understand where your taste comes from. Something interesting is, if you pay attention to the guests on *Pensado's Place,* almost universally, the ones whom I consider great had hardships as they came up.

I believe that, generally, peace, love, and singing kumbaya around campfires don't produce the fertile ground that taste comes from. I think conflict helps generate creativity and hardships in life give you something that you can't get in other ways. There's a certain case to be made that people who are happy and content may not tend to be that creative. Because it's hard to really sing the blues authentically when you have three million bucks in the bank. I'm not saying that you have to be poor and destitute and suffering to create. But what I am saying is, draw from the experiences in your life, both good and bad, and understand that your uniqueness comes from those experiences. Then reference those experiences as you create something.

WHO WANTS TO LISTEN TO THIS RECORD?

When I sit down to work on a record, one of the first things I do is try to figure out who would possibly want to buy—or stream—this record. Then I make a record for those people. I don't really care about the people who would never buy this particular recording. We'll make another record for them later. This record is for a particular group of people. To figure out how to make that record, I use the musical experiences that I've had in my life. Now, I am blessed to have had some great musical experiences. But I would argue that we all are blessed to have had great musical experiences; sometimes we just don't remember to remember them.

Church, for example, has produced a lot of great music. I don't just mean gospel or any other particular style of church music. I'm talking about all the way back to a thousand years ago. Spiritual music has had a heavy influence on how we think about music in general, and one of the things that I like about music that comes out of the church is that it has a clear purpose. That purpose is to glorify God.

I'm referencing the Western church here, but in Eastern churches—temples, pagodas, whatever you want to call them—music had, and has, a function also. There is a great deal to be learned philosophically from how music is used to glorify a deity. It's really the purest form of music, and you can use the influences of those kinds of musical experiences. If you attended religious services as a child, think about how the music made you feel. Trying to get those same kinds of feelings to come through a set of speakers is a valid way to mix.

For me, because I've played in so many bands, what I want to achieve is to get the same feeling from the music coming out of the speakers as I got playing on stage with the band. On a good night with the band, that is! But it doesn't matter. What I'm talking about here is the importance of the unique experiences you've had in your life and how you bring them into the work that you do.

Here's a cliché that will help to clarify this notion. Being able to type a hundred words a minute (which is a great skill to have!) doesn't guarantee that you are going to write a great novel. The work created on your computer comes from the life experiences you had that allow you to write the great novel. Of course, anything creative also has a technical component, like the computer, and I happen to actually enjoy that technical component. But when I sit down to work, I disassociate myself from the technical part and live in the creative part, the part that comes from experience.

Not everyone is given the same gift in terms of that balance between the technical and the creative. We wouldn't expect to drop by Bill Gates's house and see some great paintings he'd done, nor would we expect to visit Picasso's house (if he was still alive) and have him write some great code for our computers. It often takes two different types of brains to do those two different things. It's rare that you can find one person who can write serious code and also create art that moves people on an emotional level. To be able to do both, you have to fight evolution. So actually, every time you try to take a technical thing like a computer and make it do something creative, you're fighting our evolution. It's very difficult to train your brain to do both, and to do them both at the same time.

Audio people are often able to do that better than most. I've had the experience of sitting across from dozens of what I consider to be some of the most talented people in the audio space, and I'm constantly amazed at how they have the ability to be technical and creative at the same time.

In today's world, of course, because our tools have become so technical, it's now essential for producers and engineers to be both technical and creative. It's not just about mixing blue pigment and yellow pigment and putting it on a piece of canvas anymore. It's also not just about selecting a microphone and preamp and recording an artist.

I come into the studio some days and it takes me an hour just to get sound to come out of my gear. And that's the gear I use every day! There's a big technical component to our world now, but I find that's as much fun as the creative one.

ANOTHER PHILOSOPHY

I think there's a pathway for everyone and that pathways are unique for everyone. I can only interpret and share music with you from my perspective. I can't be sure how that will translate. But I know, for example, that what will translate is the music I grew up listening to in South Florida. It had a particular color to it. There was a lot of Cuban influence and a lot of Afro-Cuban influence. There was also my mom, who was into classical music and loved R&B instrumentals. My dad loved popular music from the 1940s. I didn't respect and understand my parents' music until later, when I got heavy into R&B, rock, and blues. But when I was in church as a young kid, some of the hymns would make me cry. They touched something inside of me. What's happened in my profession is, I've assembled all of those elements into my musical taste.

What I also got from all of those influences was an insatiable and very useful need to listen to as many different kinds of music as possible. Say you get a recipe book and you want to make a recipe that's in it, but that recipe is from a culture you've never experienced, with ingredients you've never tasted—well, it's doable if the recipe is written well and accurately. But to get the best results from that book, you probably need to taste that particular recipe made properly, and also to taste other foods in that particular type of cuisine, and maybe even to learn something about the culture that created that recipe. You may want to expand your knowledge of the people who created it, how they created it and why.

In the music world, we have to do the same thing. Just reading a recipe isn't really going to do you any good. If you're making a song, or an album, or a mix, you pretty much need to know what it should taste like. You also need to know the audience that's going to enjoy it. You may even want to know the history of that audience and how things got to where they are now.

Of course, you can make records, even hit records, without doing any of this. And that can be cool, and it can make a little money. But I think it's ultimately healthier to have a broader outlook on what we do. For one thing, it gives you greater fulfillment. I've had quite a few number one records and each one created a different sense of fulfillment. Some have been really fun to watch as they climbed the charts, and some I just ignored. What's the difference? For the ones that I really enjoyed watching go up the charts, it was always because of the other people involved and our relationships. We got to watch the work succeed together. We'd be calling each other every day, going, "Hey, man, we're number 80 this week!" Then, "We're number 70!"

Another thing I enjoy is knowing how much I contributed to a particular song. Some songs would have been a hit without me, and some wouldn't have been a hit unless I'd done what I did to them. The ones on which I did more were the ones I tended to like, and to enjoy and feel fulfilled by. The compensation, the money, has its place. But once the bill collectors stop calling, and you have a decent apartment and a car that runs, the extra money doesn't really enhance the quality of your life that much. What does enhance the quality of your life are your contributions to a successful record. That comes from being a part of the process, which comes from understanding the process to the point that you contribute creatively.

You could say that I'm in a unique position because I don't really create anything; I just enhance someone else's creativity. So I try to respect the original vision of the music, and to understand that vision from every possible perspective. Because communicating some of the things related to creativity—as you can tell from my contributions to this book!—is difficult to do.

For example, if we go back to our recipe and I say, "Make it spicier," that can be interpreted many different ways. Someone else might say, "Make it bigger." How do you make music bigger? Does that mean make certain elements, like the bass, stick out more? Successful communication comes from experience. Experience comes from doing. And taste, at the end of the day, is the most important commodity one can have.

MUSIC WITH PURPOSE

One last thought for you: I like music with purpose. I'm also okay with music that blatantly just wants to entertain you. I like music that has a little bit of titillation to it, music that's naughty without being vulgar. But I also like music that's a little pissed off at something. It doesn't have to be totally revolutionary, but I like when it has a point of view. I'll give an example.

The '60s generation helped end a war with music. So far, today's generation couldn't even get Occupy Wall Street going. Maybe part of the reason was that nobody was there musically for the movement. Had a group of musicians felt that the cause needed to be taken up and brought to the public through song, perhaps some needed change might have been effected.

Back in the day, the Vietnam War protests were also somewhat poorly organized. But musicians gave them a focus and a nucleus around which people could rally. People felt so strongly about something that they were able to get it onto the airwaves and into people's homes, where it helped to effect change. There was purpose to it.

There was also "We Are the World" and the Concert for Bangladesh. These were musical movements that worked to better the lives of those who were in need of help. I don't see as much of that today as I'd like to. I think part of the reason is that the music community is somewhat scattered and unfocused. The music that we're creating today has a strong entertainment factor. I support that, and, obviously, I like that. But because of my life experiences and where I came from, I wouldn't be unhappy if we had a group of musicians that got a little pissed off about things and could actually effect positive change.

Social change is a purpose, religious belief is a purpose, and of course there are some people who just need music for quality of life, and I like that, too. I work on a lot of randomly made music created just for consumption, and I try to do it well. I support it, and I like it. But I'd also like to see a little more purpose injected into the process. Just a little caring, you know. A little thought about people outside of our own sphere who could maybe use a helping hand. We can help them with the gifts we have, and it will do some good.

8 We're at the Place: Pensado's Place!

The Learning Curve

HERB: Hard to believe that in 2014 *Pensado's Place* is in its fourth year. We've filmed at three different studios, and all of them were learning labs for us. Each studio had its strengths and its challenges, and each played a role in our growth.

In our first year, the show originated from the This Week In studios, where we were "in the box." It was a tiny room staffed by a team that usually consisted of three people who ran everything, with a TriCaster module to handle graphics, lower thirds, switching, teleprompter, and everything else.

That first year was all about just getting the show up and booked. We learned what it was like to get instant feedback, how to prepare, and how critical social media was. We also got our first sponsors. Best of all, we met our producer, Will Thompson! We were grateful to have the opportunity to work at that studio, and since there wasn't a huge support team, we learned a lot about working independently. We were studious, and we maximized our learning curve. That prepared us for studio number two.

We were building a buzz and people started asking a lot of us: appearances, endorsements, messages, meetings, and the like. We were asked to be on the audio board of The Art Institute of California (AI). They had a full-on legacy television studio, and I went over and watched a production. At lunch, I ended up sitting by the program head, and by the time the first nacho was consumed, I was cutting a deal. Six months later, after rounds with lawyers and corporate headquarters, we came to an agreement and made the switch.

This was our "console" studio, full-on legacy style, with a control room, banks of monitors, a jib camera, four movables, studio lighting, floor directors, a switcher, a line producer, a teleprompter operator, and a production staff of about 15 for each show. We had about 20 cents in the bank when we started there and we spent it all on our desk for the set, which was built, literally, in an alley in Glendale, California. And we hired Will full time, which for Dave and me was scary. We were not making any real money, but, as with all growing businesses, expenses were mounting—a full-time employee, lawyers, business managers, craft services, makeup, and profit sharing with the studio. Plus we were starting to build our live appearances, starting with Gear Expo. Those events cost tens of thousands of dollars, which I had to find somewhere. And I was thinking, "Jeez, we're a Web show! When did all this happen?" The pressure was building like crazy.

What did we learn during our console phase? We learned the difference between legacy television and digital media. Their methodology, production capabilities, distribution channels, audience integration, pre- and post-production processes are different. The capture was similar; everything else was not. We learned green-room etiquette and how the pre-show hang can really set you up for a good show. Dave does the early part, holding court and creating a comfort zone for guests and visitors. That translates on air. I do a later thing where, when I arrive, it's time to take care of business. I inject a bit of urgency into everyone—including myself—that gets us focused, pumps up the adrenaline, and gets us ready to perform. Each show really is a performance, and if you aren't focused and a little bit on edge, particularly at the top, it can come off as lazy and sloppy.

We learned how to manage a disparate group of people, who in this case were students. That was interesting because each week we had folks who cared and some who didn't. That resulted in several big gaffes that Dave and I were truly pissed about. When we taped Tony Maserati, the recorder hadn't been cleared of previous material, and when it got full it stopped recording—during Tony's interview! Tony is one of the most gracious human beings on the planet, a complete class act—a close friend of Dave's and of the show. He brought his wife and was truly excited about our growth, and we failed him. I was messed up about that for months. I was determined not only to make it up to Tony, but also to minimize the chances of it happening again.

But with Alex Da Kid, another big fan of the show, as well as a starmaker and a star himself—somebody who always said yes to us—it happened again. In midstream we lost the second part of his interview and had to ask him to reshoot. He did so graciously, but I had to stop myself from going bananas. Nothing will chill your soul more than having a production team walk up to you like the Grim Reaper and give you that kind of bad news—in front of the guests, no less.

We learned about loyalty and commitment. During that period, Will produced the show, edited the show, did the social media, shot the ITL—which always required travel time—posted the show, handled guest relations, prepared notes for Dave and me, ran the students and, in some cases, the teachers also. He was amazing, and I watched him grow both as talent and as an executive. Dave and I were proud that we contributed to his growth. And then he left us! Well, not really. He got a great job offer, came to us about it, and we blessed him. Then we worked out how we could continue to work together and how his new platform as EVP of a digital agency would actually benefit the show. He'd still be involved, manage our platform and maintain his title. It's been a huge benefit with only one downside: I had to add weekly producer to my list of titles. Lord have mercy! So now I have learned those skills as well.

The last big marker for us during this phase was that we started having a ton of visitors to the set. Lawyers, TV production companies, folks from *Oprah*, PR execs, agents, managers, social media marketing types, label personnel, at-risk kids, and even sports stars. Norm Nixon, who is a former NBA star guard and current weekly television commentator in L.A., walked on our set, and when he passed by our desk, he came over to me and said, "Shit, Herb, this is like a network set." I kinda dug that!

Dave and I saw early on, though, that there were constraints working at the Art Institute that would eventually choke the show. We had a bigger vision and we knew we had to move on. We have fond memories of AI and met some fabulous students there, but we had become a commercial enterprise with expectations we needed professionals to meet.

Studio number three . . . first we had to find it! Second, Dave and I were now looking at having to pay for weekly production. That expense is relentless. I'd rather run north on a southbound freeway! We also had to somehow pull off changing studios without missing a show.

I went looking and visited several full-on television studios. After the sticker shock wore off (one place wanted 9K per week), another shock was even bigger. At five or six of the studios, they knew about us. *What?* How, I wanted to know. Turns out, mostly it was because they had had received queries from other people who wanted to create their own shows and were using *Pensado's Place* as the measuring stick. On one tour, we were shown a conference room with about 10 people working in it. We stuck our heads in quietly for a peek, and one guy looked up and said, "Oh, my gosh, it's the Larry King of audio!" Who knew?

At our current location, the same thing happened, but over the phone. Vintage King's Shevy Shovlin sent me a link, and I got the studio owner on the phone. She asked the name of the show, I answered, there was silence for about 10 seconds, and then she started yelling at people around her,

"Hey, I've got the *Pensado's Place* people on the phone!" She couldn't believe it.

We met and struck a deal, and now we are back to an in-the-box-style studio but with more production capability, a pro staff, and a cool environment. So far, so good. Our learning curve now is about telling better stories and keeping the show fresh and stimulating—also executing better, handling the financial pressure, integrating a growing team, and all the attendant things that happen when a business starts to scale. It's scary, but we're most awesomely blessed by both our audience and the audio industry. We earned it, and we continue to earn it, but we don't take any of it for granted. Not for a second.

HOW IT BEGINS

HERB: Each *Pensado's Place* episode takes up most of a week. It starts out with thinking and talking about who the appropriate guest will be. The decision is based on a number of conditions, such as who's been on lately, what's brewing for the future, who's in town, who's social media savvy, what genre of music we should cover, and whether it's Skype or in person.

Most of the time Dave makes the reach to the guest. Sometimes we may invite several people so we have contingencies. Some weeks, I'll make a suggestion, and he may tell me to run with it if I have the contact. But mostly it's Dave. His Rolodex, and his circle of friends and acquaintances, have been a huge contribution to the show.

When we have the guest confirmed we send an information package with date, location, parking instructions, contact information, etc. Meanwhile, production is ongoing. We're shooting ITLs, making sure they're edited on time, and deciding when they'll run. We also have to strike a balance with any advertising campaigns and their deliverables for that week. That means having all the assets—graphics, contest information, promo codes, etc.—delivered where they're needed so they're prepared to be inserted on show day. Overseeing that is part of the work I've inherited by taking on the role of producer; once the talent and advertising are booked I have to ensure that all of the elements get completed.

As we get closer to show day, Dave is working on the interview, thinking about music related to the guest, and figuring out his questioning approach. We also always have a pre-show call with the guest and Dave. That gives Dave a feel for where the guest wants to go, and for what that person is comfortable—or perhaps not comfortable—discussing.

We're careful about things that people aren't able to talk about. Some folks have nondisclosure agreements on certain topics, and sometimes there are things we can't discuss but can allude to. Then we sort of dance around the edges to keep it interesting!

DAVE: When you start paying attention, you notice that there are a lot of different kinds of interviews. There are guests who pretty much interview themselves, and there are some you're just not really going to get an interview out of. In that case, you've pretty much got to be both interviewer and interviewee.

Our guests mostly love talking about their profession, so once we're rolling there's always a flow. Of course, the flow rarely goes where I think it's going to, but I plan for it anyway. And usually, like a meandering river, it comes back to where I'd planned to go.

I find that the more specific and narrow the scope of the conversation, the better it works out, because that lets everyone know the boundaries. They know not to start back when they were born or with their kindergarten experiences. Or they know we're talking about a specific song.

HERB: Learning those things has been part of the process. Now the interview is usually just 20 minutes. Sometimes, though, we make decisions on air, because while we're on, I measure how it's going. If they're really rolling I may cut back other stuff. For example, we did a show with session guitarist extraordinaire Tim Pierce. Tim's interview was originally planned to be just one segment in a show about guitars. But halfway through, it was going so well that I made an executive decision to continue and have an entire Tim Pierce show—which meant that all the assets and prep stuff had to change while we were on the air!

Dave and I have gotten pretty good at working as a team. We have a private dialogue while we're on stage that allows the show to stay tight. We use notes, touches, and other little signals to each other. Since we're live and going with the moment we need to be communicating closely throughout so we don't go entirely off the rails! Of course, time is of consequence. Our editor needs to get the show down to 55 minutes. If we go too long on air, something has to suffer, and it can be tough to make those decisions. We can't hurt our advertisers, can't miss a segment, can't lose "Corner Office" because we don't want to lose our interactive audience. And we really hate to take meat out of the interview! It's a moving car at all times. Then we put it in hands that we trust to edit.

TUESDAY NIGHT; WEDNESDAY MORNING

HERB: The night before the show the production team is waiting for our notes. I need to get them all the visual assets: show title, guest, what the ITL is going to be, links, graphic pages, lots of stuff. I spend about four hours the night before laying it all out. Then I add my own teleprompter notes and the plan for the show segments.

On show day a lot happens between 6:00 a.m. and when we go live at 1:00 p.m. So by 11:00 or 11:30 Wednesday morning, the production team needs the notes and all of the assets as well as video pre-rolls, voiceovers, and any audio cues. It's also a bit tricky because they need to receive this information, including my teleprompter notes, in a way that they can clearly understand but that also fits the way I'm going to read it. If it doesn't, I'm going to have to edit it when I get to the show, and that drives me crazy. It's a version of storyboarding, and it happens every Tuesday night—and Wednesday morning, because things can change overnight.

On Wednesday morning, I arrive pretty late because I'm getting all this together. Dave gets to the show before I do, and spends time with the guests. There's that backstage comfort thing that happens with Dave for a while before the show where we try to get people to relax, and where the communication is an easy flow.

DAVE: I get there about an hour and a half early. You can—and you should, and I do—plan what you want to discuss with a guest, but if he or she goes off in an interesting direction, I don't want to be locked into something rigid. So I'll tend to have a little more material prepared than I actually need. If we go off in an unexpected direction, it helps me to have topics ready to get back to an organized flow.

I like surprises in the show, but I don't like my guests to be surprised. So I go over about 80 percent of what I think we'll be talking about in advance. An example of the benefits of that is, we had a recent guest and I'd planned on discussing three songs with him. Then, while I was going over things with him, he told me he had not done two of them. Those are good things to find out in advance! But I don't want to go too far in preparation, because I've noticed that the more I let guests talk in advance about an answer to a question, the less spontaneous they seem on camera.

I try to keep before-the-show conversation just broad enough to where it encompasses my real questions but doesn't actually contain the real questions. I also prefer that the guests don't actually answer the questions before the show. Although I have yet to have a guest who wasn't so enthusiastic that he or she wanted to answer! No matter how many times I tell them I don't really want the answers before the show, they want to reply anyway.

HERB: I'm just the opposite. I like the guest to be surprised. What I'm going to do on air, they're not going to see coming. Other than to say hello, I prefer not to talk with them until I'm on air, because I find that cool things are likely to happen when we first converse. For instance, in Young Guru's interview, I asked, "I read that you learned a lot from Jay-Z, but what did you teach Jay-Z?" It really took him aback. "Wow, nobody ever asked me that question!"

That wouldn't have worked if I'd asked him beforehand. For me it's a performance thing, and it has to come when I'm on air. So between us, we strike the balance. In Dave's 20 minutes, he might be going into a song in detail or discussing a philosophy, and he's given the guest a heads-up in advance. That makes a guest comfortable with exposing things. He or she now has an established relationship, but it still leaves room for me to add some surprises.

The questions I'm going to ask aren't ever "gotchas." If candor is required, we work closely with the guests so that they are completely comfortable and aware. As Dave does the technical, I'll take the experiential tack. Working successfully in audio requires so much more than knobs and settings; it's important to tell that story as well. The guests always tell us they had a great experience, so although the combination of our styles wasn't intended, somehow it works.

PREPARING FOR AIRTIME; SWITCHING HATS

The show requires us to switch hats in an interesting way. The directive is that, on show day, Dave and I are talent.

Don't come at us wanting to talk to an engineer or a manager or a producer. We have to perform, so if things need to be fixed, somebody else has to do it.

Meanwhile, I also have to shift gears from doing preproduction. It's handed off to the crew once I walk out of the door of my house and down to my car. I actually have to consciously think about this so that when I walk out of the door I've switched my gear. If there's a problem, somebody else has to solve it. I have to concentrate on steering the show on air, setting it up so that Dave can do his thing with all his brilliance and make the guest come through.

I can't worry about the teleprompter not working or the white balance being off, or if pixels on the camera are out of balance. All the stuff I have to worry about beforehand is complicated, but once we're rolling it's "How's the live stream going, is it pixilated, do we have enough questions in, are the questions relevant?" There are a million details that someone has to handle and we try very hard for that someone to not be me.

THE FLOW

HERB: It's always interesting for me to watch what happens in the greenroom. There's preshow prep for guests, with makeup, release forms to sign, and some sitting around waiting and hopefully getting comfortable. There are also usually a few guests of guests with us, and we've started shooting stills with our videographer, Brian Petersen, along with little "Watch me on *Pensado's Place!*" teasers.

Meanwhile, production is going on all around us and there's a checklist of elements so we're confident that all the assets are compiled. If something is missing, they'll alert us so we're prepared for things that need to be dealt with smoothly on air.

Once we're rolling, there's a lot going on: audio, sounds, streaming, and Skype. For example, if we're using Skype, Dave and I can't look at the monitor where Skype is; we have to look into the camera so that to the audience it looks like I'm talking to them. That's something you have to do a few times before you're used to it. Now we have a process where the person you're responding to is projected into the camera and you hear audio from that camera, which makes it easier, but in the beginning it was difficult for us to not talk to the monitor.

Ten minutes before airtime we bring our guests up on stage and walk them through so they know what to expect. It's just so they feel comfortable and don't have to field any curves. We want them to get a sense of the lights, camera, etc. It wouldn't be fair to people who aren't used to TV for it to be just "Boom! Let's go."

Most people, when they get to our stage, are surprised that it is full-on TV. We've had people from Telepictures, Fox Sports, Discovery, Ovation, agencies, on-air personalities—people who do television all the time, and they see no difference on our set. When they arrive, they all just say, "Oh, okay. This is TV."

WE'RE LIVE!

HERB: Once the show starts, the guests are in the greenroom while we do what we call the homework section: intro, advertiser campaigns, conversation about our week and other news. Then we run "Into the Lair," and while that's rolling we bring our guest up.

We're all at the desk for the guest walkthrough; after that Dave and I will talk together for two or three minutes, a check-in to see if we have specific things we want to bring up. Dave usually leaves the beginning to me, and I will take us where we want to go.

DAVE: The length of the opening segment is determined by a number of factors, one of which is the perceived length of the rest of the show. If it looks like it needs to be a longer show, Herb does his part, which can be very intensive. If the show were a car, that section would be the motor that keeps it going. So sometimes I don't have much to do at the beginning, other than just show my face and talk to Herb. But if something's on my mind, and I know there's time, I may talk a little bit. My part of the show tends to be a bit more improvisational. Herb's job is to get the flow going, get people introduced, and bring them in. And also, like he said, deliver all the introductory homework things, which can be extensive.

HERB: The homework section is also the commerce section. Our sponsors are watching and our audience engagement is high, so there are expectations. My thinking is always "Let's get the business done in a way that's fairly concise, and let's also make it fun." The top of the show should reflect the personalities of Dave and Herb.

I don't write the cold opening—ever—until 10 seconds before we go live. Because 9 times out of 10 I come up with something that's cooler and hipper and Pensadian just before we go live. I need it to be last-minute, because then my adrenaline gets going and it will generally trigger something interesting. So the crew will ask, "Herb, are you ready to start?" And I'll say, "Give me 10 more seconds." That's when I'm figuring out the intro.

We do try to start within a few minutes of the hour, because the live stream goes out to 500 to 1000 people. It goes to our larger platform on Thursday, but there are people who like to see the raw stuff in the initial stream. They won't see the graphics rolling in, but they put in questions for "Corner Office." Those viewers have gotten used to being a little flexible, because if there's a production challenge, some problem we need to address, we don't start until it's fixed. For example, if the Skype's not working, we don't have a show. So sometimes we may have to wait a bit.

If we have guests of guests, they can watch from our little live audience section. We've done a few shows with a full live audience, and we plan to do more. They're good for us, but it's a bit taxing on the facility because the audience has to be seated and in place for an hour. Unless you have someone managing that, it can be distracting.

While we're running, I keep track of timing. There are a number of technical ways to do this. I've done it with an earpiece, which sucks if your earpiece keeps coming out! I tend to like a monitor system, so I'm watching the time either on monitors that show the script or on other peripheral monitors. I'm always in touch with the clock. We don't generally use a floor director, because they tend to get frustrated with us! The floor director's job is basically to communicate information from the control room. But since Dave and I are managing the flow of the show from the desk on set, not from the control room, I would often either ignore them, or surreptitiously let them know I received the signal but was consciously ignoring it. That tended to get them more and more worked up, until they became like the folks who bring in your airplane to the gate, waving and pointing and generally working themselves into a frantic lather. It was actually pretty hilarious, but ultimately it was distracting. Now I just keep it all in my head and handle things from the desk.

We're a lot like the Jamaican Bobsled Team. We don't make hard left or right turns; we kind of ease into things. If we have something good going, I don't want to stop it. But I still have to keep things moving and make sure we're on track with the timing.

While we run the ITL, the guest comes back to the desk and they check audio. While we're counting down from the ITL, I'll ask Dave what he's going to come back with; I introduce him referencing that, and then we go right in.

We've worked on being concise and to the point. So once I've introduced, and Dave starts questions, we just tag team. I stay out of the way a lot. There have been shows where the chemistry is wonderful, Dave is hitting home runs in a row, and I may not come back in at all. There are other shows where I'll wait, then take the conversation to another place, which allows Dave to go to yet another place to come back for follow-ups. In general, I like to leave that first 15 minutes to Dave and the guest. Breaking down various pieces of music, or discussing the philosophy behind how they work, just doing what Dave does, he's generally excellent.

DAVE: I don't think about any of that stuff! I don't have a feel for the show while it's going on. It's a good thing Herb is great at that. Without him I might easily do an eight-hour show and not even know it. Herb should be commended because he ensures that won't be inflicted upon the world. It could be the most boring thing on earth!

I have a tendency to like to ask long questions, so in my head I'm usually trying to edit myself. I spend the most time deciding what the first question will be. The importance of the first question is that it can determine the direction the show will go and the holes that are left for Herb.

A good tight first question also makes everything better because guests tend not to know the proper length for an answer. If I ask a question that could be answered by a novel, I'm probably going to get a novel. So I try to keep the first question somewhat yes or no. Also, some guests are a more nervous than others, and I don't want to ask the best questions when they're a little uptight. If I notice that, I may want to go to lighter stuff, and then ease back in. It's kind of a bob and weave. But all of this I'm describing here doesn't really enter my mind as thoughts.

The first six months I think I was more concerned with how I was perceived, and that my questions would help me look good! Now I don't really care how I'm perceived—well, I do, a little—but I try to make everything more about the guest.

When we first started, I felt like I had one show to get in my entire life's knowledge and information. I felt the same about our guests. Now I know the guest will most likely be returning, so I don't have to cover everything the guest has done. I can focus on being in the moment and remove myself somewhat from the process. My favorite moments are those little happy accidents when I learn something that I didn't know about that guest, something that catches me off guard. Those pearls resonate on the Internet and I love those moments. You can't plan for them, but hopefully you can create an atmosphere where they come to fruition.

I try to come up with questions that the audience will want to know the answer to. Everything about *Pensado's Place* is designed to be the best it can be for the audience, so we put the audience first in every way. I also try to enjoy myself while I'm there. I tend to feel that if I'm entertained, the audience will be entertained. I'm not a professional at this, but I'm getting better all the time, and there's always something I want to learn from each guest. We had a guest who was a composer for film and TV, an area I know nothing about. So my questions were more in the spirit of investigation than when I have one of my good friends on who's just had a number one record. Everything varies. I guess you could say Herb's role is all about consistency and mine's all about inconsistencies, but hopefully planned and controlled inconsistencies. My early inconsistencies, when we first started, didn't always lend themselves to making a better show, but I've learned to be more aware of all of the elements. If you're not, there are many small things that, combined, could make the show less than entertaining.

HERB: The way you get to those happy accidents and inconsistencies that make for magic is to set up a consistent background so that everything works. That leaves room for the good, spontaneous stuff to happen.

We've had shows where things are going wrong around us, and there is nothing more terrifying than being in the middle of the show, looking over into the control room, and seeing people panicking.

I pay attention to everything. I guess I could just be looking at the camera and the guests and ignoring the control room, but I'm not designed that way. I need to know what's going on. And when you've experienced a few situations where someone comes out of the control room and tells us we've lost the whole show, it's most definitely not fun.

If you have to do things over, you are going to lose energy. And energy is what connects on TV. So the guest has been at the studio two and a half hours, and he's been great. Then I see scrambling in the control room, and someone walks out and says, "Uh, Herb. . . ." We have to get the talent to do it over, and hope we can re-create the energy. But you're not going to have the same thing. Now, of course the audience didn't see the first interview, so for them it might be great, but for us, we're doing the on-air traveling-circus balance thing.

So we do the interview, we hope there are no problems, and we go through the seconds: "Batter's Box" and "Corner Office." Then the fun part: Dave and I have to coordinate how to end the show. Because "Corner Office" may have five relevant questions and it may have two. The host of "Corner Office" is sending me signals—two more, one more, whatever. At that point we have Dave wrap up and we end with a standard pattern. We thank the guest and ask for his or her commitment to things we're going to do in the future. They always say yes.

DAVE: One hundred percent. That's the final push. If they were close before, being on the show is the final motivation. People have no idea how much effort, time, personnel, and expense goes into the show, so they generally come expecting something less than what we are. We're full-on TV, like Herb says, and when they realize that, they also recognize that anything we do is going to have a unique character and quality and be the best that it can be. They see the future of where we're going and they want to be a part of it.

HERB: Finally, it's "Dave, take them home." Usually in that last moment he'll find a philosophical point of view, a teaching point, some way to land on some really cool Dave thing. It's a curlicue at the end of the show that's generally really special. To me, that signature at the end is a summary of the authenticity of the show. It's where Dave connects with people about what just happened and promises to be back next week. It's an important moment.

Immediately following the show, we shoot some bits with our guests for the "Recall" segment. That's an extra four or five minutes of additional footage, things we didn't have time to get to. Then we take photos on the set, and there's almost always time spent with people who are in the audience and want to come up and talk. So there's that 10 or 15 minutes post show on the set where people are asking questions or taking pictures and everybody decompresses. The guests are not usually in a hurry to leave!

DAVE: There's always that magical moment when we go, "It's over? That went by so fast!"

It still does go by really fast for me, but it's also kind of cute because every week the guests all say the same thing: "That was an hour?"

HERB: After we've wrapped, Dave and I go our separate ways, but we usually connect by phone within an hour to touch base and review the show.

POST-SHOW POSTMORTEM

DAVE: It's a long ride home for me, and when we both get on the 405, I'll call Herb. That's when I can find out, not only what I did badly, but what I did well. Because sometimes I did good things and didn't realize it.

I'm still learning, so there's always an opportunity for me to get better. I believe that if you want to do something in life, your goal shouldn't be just to get it done. Your goal should be to do it the best you possibly can. And I want to be good at doing this.

HERB: We don't have the luxury of not being good at this!

DAVE: I noticed that there was a need to get better at about the halfway point that first year. I was still recovering, and we were still meeting at the chicken place and riding to the studio together. The conversations on those early rides home were all about how I could improve. Now, you might read this and think, well, didn't Herb need to improve? But the way the show is constructed, I'm probably on camera a little more than Herb. He's not on camera as much and he's already good. So it's my responsibility to catch up to him. And hopefully to surpass him—in a friendly way.

Those rides home, and those conversations after the show, are still how we decompress and talk about the moments that came along. Herb's very good at measuring his abilities. I'm not gifted in that arena. In my day job, I'm pretty good at knowing when I've done a good mix. But in the show world I need help figuring it out.

"Problems don't age well."

HERB: I get information from a lot of sources. I hear from the sponsors and the producer. I hear things before and immediately after. Then, of course, we have the comments that come in.

My belief is that if everything else isn't going well, we're not going to get a chance to be brilliant. That's why I work so hard at getting the details right. We're going to look stupid if the machinery looks bad. I include myself in that process; I'm measuring myself while I'm on air. How I appear, what words I hit, what I'm emphasizing. What works this way, what sets that up. I believe you have to take all of this very seriously.

EDIT AND POST

When Dave and I are done and heading home, there's still a whole process happening back at the studio, where they're getting the show that we just taped into a file and into an editor's hands. The editor goes through the process of putting in lower thirds, graphics, etc., and also does some tightening and cleanup.

I try hard to stay out of the way of the people who work on our behalf. I don't want to have to hold their hands, so I try not to get involved with editing. Dave and I are so close to it, we may be protective of things that we shouldn't. Instead, I trust their editing genius and hope for a fresh point of view that Dave and I may not have. So once we finish the show we don't see it again until it's edited and posted.

That process also includes preparing it for the various uploads. You have to upload differently for Apple and iTunes then you do for mobile or YouTube. An easy show takes two to three hours of prep, a hard show longer. We need to post the show at 2:00 p.m. on Thursday. If we miss, the audience gets uncomfortable.

Our viewers are very vocal; they have expectations. If we post late, we will definitely hear from them. "Where's the live stream, where's the show, where's the answer to the Vintage King question? You said it would be here, it's not there!" They rely on us to be consistent.

If the production team does have a miss—and nobody's going to be perfect—we just make sure that we correct it. The cool part is that we can talk to the audience, let them know what happened and where they can find what they need. The platform management people will inform us if something needs to be corrected, and they'll communicate that to the audience: "This will be posted Friday at 2:00 p.m. on all the platforms," or whatever the message needs to be.

The show is seen as much on YouTube as it is heard on iTunes. We stream on Justin.tv, and we have segments, like "Into the Lair" and "Batter's Box," that are posted and watched separately. It takes time to get all of that right, and our team does that dance weekly— for viewers in 187 countries!

The last thing we do is look at comments and views. Most of that comes in over the weekend. We have people who measure all of it in terms of what it means statistically. I tend not to look at that, because I don't want my week to start up or down based on a comment. But once we hear some of that statistical information, Dave and I will talk about what it means. By Monday we have a pretty good read on how the last show did.

There's other communication with our fans during the week, of course. There's the Facebook page, e-mail, people who tweet, and we have to decide if we need to tweak something or answer questions. People are often shocked when we get back to them personally, but we've created a one-to-one relationship with people mostly because we're accessible. I try to get back to them quickly, in part because our fans tend to be very grateful for the responses, but also because a lot comes at me. Replying quickly gets stuff off my plate. It's turned out to be good practice, and it also helps the show.

So we're edited, we're posted, we're checking feedback, and we're done! Then the circus starts all over. We don't get any time off. We're on the Internet with a show every week. We can build in hiatuses as long as we have content, so sometimes we have to triple-book shows. But that's the rodeo! Onward!

9 Gifts from the Gods

Some of Dave's favorite techniques that our *Pensado's Place* guests have shared on the show

DAVE: One of the cool things about the show is that every week I get to sit across from some of the most gifted and talented people in the world of audio. And every week I learn something new! I've been doing this more than a minute, so that's pretty surprising to me. I think I know a lot, but then I sit across from these guys and realize I really don't know much at all!

With that in mind, I'm going to share with you some things from the show that particularly caught my ear. Since everyone's learning level is a little bit different, each of you will probably take something different away from these items. But they are all things I'm now using on a daily basis. Some are things I'd forgotten, some I never knew, and some I actually taught to people who are now teaching them back to me again!

AL SCHMITT

Let's start with Al Schmitt, because, of course, we all stand on Al's shoulders. In Episode 117, Al described how, when he's working at Capitol Studios in Hollywood, he takes feeds from their five classic live reverb chambers and brings each of them back onto his console in mono. One is panned hard left, one hard right, and one straight up the middle. The remaining two split the difference between dead center and hard left and right.

I thought that was a cool idea, so I tried it with plug-ins. Instead of using five different mono reverb plug-ins, I went with just three and panned them left, center, and right. Left and right were individual mono plug-ins, but they were very similar sounding. The center plug-in was completely different. Then I panned to various amounts of those reverbs with different elements. That turned out very usable and I now include this technique on almost every mix. Keep in mind that you don't have to use impulse responses that are chambers; you can use anything. The most important part of the concept is that they're mono, and that you don't necessarily send everything to all three. Sometimes you might send a guitar panned right to just the left reverb return.

JACK JOSEPH PUIG

Number two is from Jack Joseph Puig (Episodes 22, 91, and 92). This is something I'm still thinking about. Jack said that he uses compression as EQ. I haven't worked this all out yet, but it's haunting me. He's explained it a couple of times and I still don't quite get it, but apparently it has to do with how you manipulate the attack, the release, and the amount of input. Depending on where the threshold ends up, you can alter the sound and also the timing. I'm still struggling with this, but I love the concept! Let me know what you find out.

JUSTIN NIEBANK

In Episode 75, Justin Niebank discussed spring reverbs. Several guests, including Andrew Scheps in Episode 73, have mentioned spring reverbs. I'd never spent much time with them, because when I first started engineering we had one and it didn't sound very good. I don't think it was working correctly, and that turned me off to it. Plus, as a guitar player, I always thought spring reverbs were just something that came with Fender amplifiers! But now I'm using mono spring reverbs on everything. I particularly like them on vocals, and I appreciate that Justin brought this to my attention.

MANNY MARROQUIN

I'd never thought about having more than one stereo bus, but Manny Marroquin (Episodes 6 and 105) described having two different stereo bus compressors on the same mix and using them in different ways, not necessarily at the same time. That led me down some paths that I feel really improved what I do.

GREG WELLS

Greg Wells (Episodes 34 and 93) talked about advice he got from Jack Joseph regarding comparing your mixes to other mixes while you're going along. This not only tells you when you're doing something that's not up to par, but also lets you know when you've done something above par. It's something I've been doing for a while, but having engineers of Greg and Jack's stature say that they do it too made me feel good.

CHRIS LORD-ALGE

Chris Lord-Alge (Episode 80) said that spending too much time on the mix process may actually destroy the mix, because it can destroy your creativity. Spending too much time is sometimes not in your favor. That's something I struggle with every mix. I take way too long!

TOM ELMHIRST

Tom Elmhirst (Episode 83) talked about reducing the number of tracks that he's given to a more manageable number so that his creativity can be put to use mixing broader parts of the song, as opposed to individual background or guitar parts.

He likes to do sub-mixes so that he ends up with only about 20 pieces of information. I started doing that and I really liked it. Chris Lord-Alge also mentioned that he believes in "less is more." I think this is a good point, and I work that way a lot.

MICHAEL BRAUER

Michael Brauer (Episodes 18 and 70) made a couple of philosophical points that I really liked. He said that he'd rather be the guy who came up with new ideas than the guy that chased other people's new ideas. Thinking about that inspired me to push myself a little harder.

A big chunk of my work is based on Brauerizing, which is how Michael Brauer uses multiple stereo buses and multiple sends. Of course, this concept has morphed a little bit for him since the days when he was working all analog, now that he's using Pro Tools. But I've modified my system somewhat to where I use four or five aux sends. For every mix, every track goes to one of those four or five auxes and those auxes feed the stereo bus. I'll have a drum aux, a vocal aux, and an all-music aux. I also usually have one for bass, and sometimes for effects and other groups. With drums, for example, it allows me to get them under control a little bit. I can almost treat them like the song is a just a solo drum part. I can affect the drums before I feed them to the stereo bus, so the stereo bus compression and limiting don't have to work as hard. For another example, with vocals or keyboards, sometimes you may want to make them a little brighter but you don't want the drums a little brighter. With separate auxes you have options. You can raise group levels, play with those auxes, and somewhat reconstruct your mix without changing a lot of individual instrument levels. You can see how it feels with the vocals a little louder, the keys a little louder or softer. Brauerizing! It works its way into every mix I do.

DYLAN DRESDOW

Dylan Dresdow (Episodes 3 and 122) pointed out that due to the proliferation of technology we're now able to reduce the number of initial choices we need to make—in a good way! As an example of how he did that on a recent Black Eyed Peas project, he mentioned what he calls working in a nonlinear fashion. He did the mix and then sent it to mastering, where the mastering guy did his thing and then sent it back. Then will.i.am did his further overdubs based on what he heard in that mastered version. Dylan could also make changes based on what he heard in the mastered mix, like adding more transients if they were getting knocked down a little too much by the limiting. That's a very cool use of modern technology that I take advantage of.

Dylan also mentioned that for will's solo project he set up live speakers in the studio along with smoke machines, lights, and lasers in order to recreate the fun of a live experience. He'd go out in the live room and listen to the mix coming out of the speakers with the lights and smoke while he was controlling it from a laptop or an iPad.

That made me recall that there have been several songs I've mixed where I fed the drums back out into a live room, mic'd the room, and then fed that ambience back into the drum sound. I need to do that again!

JAYCEN JOSHUA

Jaycen Joshua (Episodes 8 and 127) talked about how he likes to incorporate a lot of different effects—like 40!—

into the template that he uses to start his mixes. He uses the template because it takes some of the thinking and technology out of the equation, allowing him to be more creative. With a basic concept worked out, you can tweak and tighten to fit the song and make it unique. A template takes a lot of the heavy lifting out of the way when you're first getting going, and I've been using that technique a lot.

Jaycen also explained what he did with some background vocals that I liked on a Mike Will song. The top end on them was especially nice, and he said he'd used Waves' LoFi or SoundToys' Decapitator on them because he was trying to separate the sound of the backgrounds from the sound of the lead. He suggested that you should probably do that early on in the mix so you're able to mix around that particular sound and make it work within the overall concept of the mix.

Jaycen also likes to use the Imager section of Izotope's Ozone 4 and Ozone 5 so that instead of widening the whole song, he can just widen the mid range. And finally, he shared what he called the "tip of the century." Sometimes when he takes a bass drum and makes the low end as massive as he wants it, it will fight the bass. He found a trick that the EDM guys were using a lot but that hadn't really found its way into the hip-hop and pop worlds yet.

When he sets up the bass he splits the frequencies over three tracks, so he has high-end, mid-range, and super low-range tracks of the bass. Then he'll side-chain just the low-end track of the bass. That's what distinguishes this tip from the way a lot of the EDM guys use it.

JIMMY DOUGLASS

Jimmy Douglass (Episode 140) said that when he's changing genres he likes to immerse himself in the new genre, both to clear his head and to help him keep up with what's going on. In general, of course, it's a good idea to listen to as much music as possible—all types of music. It helps you develop the ability not only to follow trends, but also to predict them. I thought that was inspirational. He also mentioned the "Jimmy Douglass Auto-Tune Technique." Everybody should go to Episode 140 and check this out. It's pretty clever.

JASON SUECOF

Jason Suecof (Episode 135) said the Kemper Profiler amp had changed his life. He called it the coolest guitar invention in the last 20 years, and I have to agree. I think the Kemper Profiler amp is one of the best boxes to come along in years. I'm about to get one!

TONY MASERATI

I was really impressed with the description that Tony Maserati (Episode 126) provided about the kick drum on one of the Pink songs that he mixed. I liked his thought process regarding how he EQ'd it and how he determined the space it should occupy in the mix. It was enlightening to hear the decisions he made about something seemingly so simple: a kick drum and how it hits you, how it envelops you, and how the low end should surround you. I've tried to incorporate some of that into my work and I'm really happy with the results.

Tony likes to envision an imaginary room that the mix is occupying—it can be a concert hall, a party, anything. On "Blurred Lines," one of the reasons he put some of the parts on the right was that he wanted it to sound like Pharrell and Robin Thicke were actually at a party. The lead vocal's in the center, but there's a party going on behind the lead vocal. He panned the cowbell over to where the party was, and he selected EQ and effects to amplify the party atmosphere. I really liked the idea of having that kind of imagery for the mix. He also mentioned a Pendulum limiter that he's been using since hearing about it from Michael Brauer. I checked it out and I liked that a lot, too.

RODNEY JERKINS

I had commented to Rodney Jerkins (Episode 114) that I felt like the bridges in his songs tended to be very special. It's obvious he puts a lot of time and care into creating them. He explained that he likes the bridge to be just that: a musical passage from one part of the song to the end of the song, and he likes to create tension and release within it. He also always tries to add something new after the bridge, which sometimes means the vocals need to get a little louder. I tend to make my vocals get louder as the song goes on, but it was cool the way Rodney described the purpose of that new sound, what it needed to do when it came in after the bridge and how the vocals needed to support that. I've expanded this a bit and tried to make sure that I'm working to extract the unique purpose of each section of a song.

DEREK ALI

Derek Ali (Episode 97) discussed using filters for enhancement. I think he and I got a little confused and erroneously mixed up low-pass and high-pass filters! For what he was describing, he wants the low end to pass through. He rolls the top end off a lot of sounds, and I asked him why. Turns out it's because, when he first started out, he had a very limited number of tools, and he found filtering was the easiest way for him to get the sounds he wanted. Filtering is now part of his style, and if you listen to the Kendrick Lamar record he did, you'll hear a really masterful manipulation of the high end. Since that conversation, I've been playing with some of that style for my mixes and I'm very happy with it. Derek also mentioned that he uses a plug-in called Smack, by Avid, on his hi-hats. I'd never thought of that! I'd used Smack, but just to tame something. I tried it on a hi-hat and really liked it.

SUMMARY

To round all this out, what I'd like for people to take away from this chapter is the understanding that when you hear these great engineers talk about their work, what's more important than the techniques and tools they use is the insight into their thought processes. Techniques apply to specific situations in specific songs, but what applies to every song is the way these engineers thought things through, and why they decided they needed to create or alter a technique. It's the reasons they did certain things that are most important. What I really like is learning how they think. When you understand that, you can apply their concepts to the way you do things. That's the real beauty of all of this.

AFTERWORD

We began this book with a call—a miraculous call—from a hospital. Little did Dave and I know that it was the first step in a march of miracles.

From our weekly meetings with brilliant people, to realizing unimaginable opportunity, to becoming global educators, to discovering the off-the-charts passion of the very best audience in the world, to the incredible support of our professional peers and the spotlight of celebrity, we stand humbled and amazed at the responsibility that's been handed to us. We take it very seriously, because it's very important to us—but it's not what's most important.

The journey has not been all roses. There was personal financial risk, and many foes had to be vanquished. There were many stresses and obstacles to growth, and there was a measurable toll on family and friends—but that's not what's most important, either.

What's truly important—truly at the heart of "our" thing—goes back to that first call. My boy is back, healthy, brilliant, and killing it! It's as simple as that. We've journeyed from the precipice to the pinnacle, always holding to our core principles of unshakeable loyalty and brotherhood. Yes, Dave feels the same way about me—he says it constantly—and he looks out for my well-being as fiercely as I do his. The fact that we're building something of relevance based on those principles gets us going every single day.

We pray that you find "your" thing like that—something special where passion is your driving force. It ain't easy to find, but when you do, look out. It'll take you places you can't imagine. Search for it.

Finally, please stay with us. As we look to the vistas of the future, we want you there with us. It's no fun to go it alone. To our private warriors—the PTSD, disabled, sight-impaired, sick, and temporarily down-on-their-luck folks who face daily hardship—please continue to reach out. Use us to help you hold on—to fight to see another day. Take a break, let us make you laugh, let us help you learn, or just come see us in person. Know that *Pensado's Place* is always here for you. Always.

And as a group I once worked with famously sang, "Come along and ride on this fantastic voyage!" The journey continues.

Dave and I love you.

—Herb

INDEX

Page numbers in italics refer to photographs.